Microbial Food Poisoning

Second Edition

Edited by

Adrian R. Eley

Department of Medical Microbiology,
University of Sheffield Medical School,
Sheffield, UK

CHAPMAN & HALL

London · Glasgow · Weinheim · New York · Tokyo · Melbourne · Madras

Published by Chapman & Hall, 2–6 Boundary Row, London SE1 8HN, UK

Chapman & Hall, 2–6 Boundary Row, London SE1 8HN, UK

Blackie Academic & Professional, Wester Cleddens Road, Bishopbriggs, Glasgow G64 2NZ, UK

Chapman & Hall GmbH, Pappelallee 3, 69469 Weinheim, Germany

Chapman & Hall USA, 115 Fifth Avenue, New York, NY 10003, USA

Chapman & Hall Japan, ITP-Japan, Kyowa Building, 3F, 2-2-1 Hirakawacho, Chiyoda-ku, Tokyo 102, Japan

Chapman & Hall Australia, 102 Dodds Street, South Melbourne, Victoria 3205, Australia

Chapman & Hall India, R. Seshadri, 32 Second Main Road, CIT East, Madras 600 035, India

First edition 1992
Reprinted 1994
Second edition 1996

© 1992, 1996 Adrian R. Eley

Typeset in 10/12pt Berkeley Old Style by WestKey Ltd., Falmouth, Cornwall
Printed in Great Britain by St Edmundsbury Press, Bury St Edmunds, Suffolk

ISBN 0 412 64430 4

A catalogue record for this book is available from the British Library

Library of Congress Catalog Card Number: 95–72202

♾ Printed on permanent acid-free text paper, manufactured in accordance with ANSI/NISO Z39.48-1992 and ANSI/NISO Z39.48-1984 (Permanence of Paper).

Contents

Contributors

A.R. Eley, MIBiol, MSc (Ed. Man), PhD, Cert. Ed (Inf. Tech), CBiol, FIBMS
Senior Lecturer, Department of Medical Microbiology, University of Sheffield
Medical School, Beech Hill Road, Sheffield, S10 2RX

I. Fisher, MCIEH, FRSH, MRIPHH, MIIRSM
Environmental Health Consultant, Grove House, Ripley, Derbyshire, DE5 9TD

M.O. Moss, BSc, PhD, DIC, ARCS
Senior Lecturer, School of Biological Sciences, University of Surrey, Guildford,
Surrey, GU2 5XH

T.A. Roberts, BPharm, MA, PhD, FIFST OBE
Former Head of Microbiology Department, Institute of Food Research, Earley
Gate, Whiteknights Road, Reading, RG6 6BZ

J.C.M. Sharp, MBChB, DPH, FFPHM, MCRP
Former Consultant Epidemiologist, Scottish Centre for Infection and Environ-
mental Health, Ruchill Hospital, Glasgow, G20 9NB, and Honorary Lecturer,
Department of Infectious Diseases and Public Health Medicine, University of
Glasgow

Preface

Yet his meat in his bowels is turned, it is the gall of asps within him.
He hath swallowed down riches, and he shall vomit them up again.

Job 20: 14–15

Over the past few years, food poisoning and food safety have become very topical subjects, eliciting a great deal of public concern both in the UK and elsewhere. During tutorial sessions with medical students in the late 1980s, I found myself being asked to recommend appropriate textbooks on food poisoning. At that time, I had to admit that there were few books available on this topic, and none which I felt was designed to meet their particular needs. This was the initial stimulus which prompted me to produce this book.

Microbial Food Poisoning was never intended to be an authoritative work of reference on the topic: it began life as a teaching aid for senior medical students in the UK, which aimed to cover the major aspects of the subject in sufficient detail to be instructive without being confusing. The finished book has a rather more international flavour, using examples from overseas wherever relevant. It is also, perhaps, somewhat more broadly based, and as such should also prove to be of interest to students of microbiology, food science and food technology, to professionals allied to medicine such as nurses and medical laboratory scientific officers, and to environmental health officers and catering staff. In summarizing what is known about established agents of food poisoning, I have tried to emphasize the modes of pathogenesis of organisms; I have also included emerging pathogens and examples of newer technologies, in order to indicate where future developments in this field may lie.

One of the things which struck me most forcibly when preparing this book was the fact that there is, as yet, no single universally accepted definition of the term 'food poisoning'. I should like to make it clear that the working definition I adopted for the purposes of writing this book is entirely my own, and has no pretensions to universality. What the definition is, and the reasoning behind it, is explained in Chapter 1. May I end with the hope that those who read this book will find that it promotes increased awareness and discussion of an important topic, and also that they never have direct personal experience of its subject matter.

A.R. Eley

Preface to the second edition

As this second edition of *Microbial Food Poisoning* goes to press, food poisoning and food safety are every bit as topical as they were when the original version of the book was in preparation. The incidence of food poisoning continues to rise in the UK, and we have also witnessed, in the USA especially, an increase in so-called 'emerging infections', many of which have been food-borne. The need for greater awareness of the subject is still as pressing as it was in 1991.

One aspect of the situation that has changed since the first edition is that a broad definition of the term 'food poisoning' has now been adopted in the UK, and this should prove helpful in notification and recording of the disease. I have, however, still retained my own working definition of the term, which has allowed me to structure my discussion in what I hope is a clear and comprehensible way.

Although the aims of this revised text are little changed from the first edition, I have been grateful to readers for many helpful comments and now include a number of modifications designed to improve coverage and bring it up to date. Our understanding of food poisoning continues to develop: recently a new *Helicobacter* species has been described as a possible new cause of gastroenteritis. There have also been advances in laboratory diagnosis and in the use of live avirulent strains of salmonella which show promise in controlling carriage of the organisms in chickens. By including details of such developments I hope to prolong the useful life of this text for a broad spectrum of potential readers.

A.R. Eley

The authors and publisher accept no responsibility for the treatment recommendations in this book.

Acknowledgements

Many people have had a hand in the preparation of this book; I should like to take this opportunity to express my grateful thanks to all of them, and particularly to the following individuals: my four co-authors, Mr I. Fisher, Dr M.O. Moss, Dr T.A. Roberts and Dr J.C.M. Sharp for their valuable contributions in their specialist areas; my colleagues Dr Mark Wilcox, for finding time to read and comment on the entire text and advise on clinical matters, including chemotherapy, Mrs Hazel Storer, for typing the manuscript so swiftly and efficiently, and Mr Ian Geary, for drawing the figures and designing the book cover; Professor C.R. Madeley of the University of Newcastle, for providing the electronmicrographs, Dr D. Grundy of the University of Sheffield for advice on pathophysiology, Mr P.N. Sockett of the Communicable Disease Surveillance Centre, London, for food poisoning data, Mr W. Hyde of Lab M, Bury, for technical advice on laboratory media, Mr R. Hart of Sheffield Health Authority for information on the Food Safety Act 1990, Dr Naeem Akhtar and Mr Kevin Oxley, both of the University of Sheffield, for providing photographs of DNA fingerprinting and bacterial plasmids respectively; Mr Nigel Balmforth of Chapman & Hall, for being a source of constant encouragement and optimism; and my wife, Dr Penny Eley, without whose support, encouragement and editorial help this book would not have been written.

I should also like to thank the following for their permission to reproduce copyright material: The Editor, *Communicable Disease Report*, Public Health Laboratory Service (Tables 1.2, 1.3, 2.4 and Figure 4.1); The Editor, *The Lancet* (Figures 1.3, 2.1 and 6.4); The Ministry of Agriculture, Fisheries and Food, Economics and Statistics (Food) Division (Figure 1.3); CRC Press Inc., Boca Raton, Florida, USA (Figure 6.4); Wrightson Biomedical Publishing Ltd (Figure 7.3); *British Food Journal* (Figures 7.1 and 7.2); Professor W.M. Waites, University of Nottingham (Figure 1.3); Dr A.C. Baird-Parker, Unilever Research, Bedford (Figure 2.1); Dr Casemore, Public Health Laboratory, Glan Clwyd Hospital, Clwyd (Figure 6.4).

A.R. Eley

Acknowledgements for the second edition

In addition to my acknowledgements for the First Edition I would again like to thank my four co-authors, Mr I. Fisher, Dr M.O. Moss, Dr T.A. Roberts and Dr J.C.M. Sharp for their valuable contributions in their specialist areas. I have been fortunate to have had the benefit of the knowledge and experience of Drs Roberts and Sharp and I wish them well in their retirement. I would also like to thank the following: my friend and colleague Dr Peter Cowling for reading the entire text and advising on clinical matters including chemotherapy, Ms Pauline Whitaker and Mrs Gillian Griffiths for their word processing skills, Mr Ian Geary for drawing new figures, Mr Kevin Bennett for providing a photograph of ribotyping, Dr J. Cowden of the Communicable Disease Surveillance Centre, London, for food poisoning data, Dr D. Jones of the University of Leicester for advice on *Listeria*, Dr Peter Chapman of the Public Health Laboratory, Sheffield for information on VTEC, Mr Nigel Balmforth and colleagues of Chapman & Hall for encouragement and my wife, Dr Penny Eley for her support, and editorial help.

I should also like to thank the following for their permission to reproduce copyright material: Marcel Dekker (Table 3.6); Wrightson Biomedical Publishing Ltd (Table 3.7); MCB Publications Ltd (Figures 7.4, 11.1 and 11.2); Highfield Publications (Table 11.3).

A.R. Eley

1 Introduction

A.R. Eley

1.1 WHAT IS FOOD POISONING?

A wide variety of diseases can be caused by eating food contaminated with pathogenic micro-organisms or their products; by no means all of these diseases can be classed as food poisoning. The question of which pathogens associated with food should be seen as agents of food poisoning can be considered from a number of different angles, and it is clear from the literature on the subject that, despite a measure of consensus over specific organisms, a range of different answers is possible, depending on the definition of food poisoning adopted. Before 1992 different definitions were in use within the UK. However, in late 1992 and primarily for the purpose of notification, the Chief Medical Officer recommended that the following definition be adopted: 'Any disease of an infectious or toxic nature caused by or thought to be caused by the consumption of food or water'; this has since been adopted by the World Health Organization. Although we approve of this broad definition, for the purposes of this book we wanted to concentrate on those micro-organisms mostly affiliated with food and which result in disease. Therefore for this reason, we feel that it is appropriate to indicate at the outset what our approach to the term 'food poisoning' has been.

We have been guided by two basic principles: firstly, the notion of 'poisoning' itself, and secondly, the distinction between food poisoning on the one hand and food-borne and food related diseases on the other. The shorter *Oxford English Dictionary* defines a poison as 'a substance which destroys life by rapid action and when taken in a small quantity'. Only a small proportion of the conditions with which we are concerned are life-threatening in normal patients, but almost all are characterized by 'rapid action' – that is, the rapid onset of symptoms (typically nausea, vomiting, abdominal pain and diarrhoea) after ingestion of contaminated food. As our title indicates, we deal principally with 'poisons' of microbial origin; in order to give a broad view of the general subject, we include agents such as viruses, mycotoxic fungi and protozoa which can be responsible for similar conditions to those caused by bacteria when present in food. We have specifically omitted organisms such as *Salmonella typhi* and the hepatitis A virus that produce characteristic diseases which usually have quite long incubation periods. In the case of such named diseases, patients generally present with distinctive clinical features rather than with gastrointestinal symptoms.

The majority of the organisms which we consider as agents of food poisoning are also capable of rapid multiplication and/or manufacture of toxins in many different foodstuffs (the exceptions being viruses and protozoa). In this respect they can be distinguished from the causes of diseases such as brucellosis, where food is only the vehicle of infection, and is not able to support growth. We do not deal with the majority of these food-borne diseases. Nor do we dwell on food-related diseases, which can be defined as certain chronic conditions of an immunological – as opposed to a gastroenteric – nature, caused either directly or indirectly by food-borne pathogens. In such conditions, micro-organisms in food can act as 'environmental triggers'. Autoimmune disease caused by tissue damage in organ systems may follow the interaction of the organism or its products with the host immune system. Alternatively, immunological deficits or autoimmune disease may be due to disrupted nutrient transport caused by an organism or its products interfering with intestinal integrity. Food-related diseases include septic and aseptic arthritis, Graves' disease, Guillain–Barré syndrome, myocarditis, IgA glomerulonephritis and haemolytic uraemic syndrome (HUS).

In the last few years there has been publicity concerning specific food allergies some of which require hospitalization and more rarely lead to fatalities. These unpleasant reactions to foods are a complicated puzzle but almost certainly are unrelated to micro-organisms or their activity.

Finally, mention must be made of bovine spongiform encephalopathy (BSE), a disease of cattle similar to scrapie in sheep, which is probably caused by a so-called 'slow virus' or prion. Since it was first recorded in the UK in 1986 more than 100 000 cases have been reported. And, although figures are now declining there is a distinct lack of knowledge of the incubation period in humans, so that if the agent could be transmitted to humans the scale of the disease may not be apparent for a considerable period of time. Despite its topicality, and concern over the possibility of transmission from cows to humans via infected meat or milk, we do not discuss BSE in any further details, since it has not yet been identified as a disease of humans, and appears in cattle to have far too long an incubation period (approximately 5 years) to conform to our notion of 'rapid action' as being a major characteristic of food poisoning.

By our definition, therefore, microbial food poisoning is an acute condition, usually presenting as gastroenteritis, the first symptoms of which normally arise within a few hours or a few days of consumption of food containing pathogenic micro-organisms and/or their products. In most cases (the exceptions being viral and protozoal infections) the food in question will have supported rapid growth of pathogens. The mycotoxic fungi are capable of growing on a variety of foodstuffs, and may be responsible for acute gastrointestinal disease. They are also associated with longer-term poisoning, their toxins giving rise to a range of chronic conditions, including cancer.

1.2 TYPES OF FOOD POISONING

There are many different types of food poisoning, of both microbial and non-microbial origin. Table 1.1 gives an overview of the various possible causes. The vast majority of reported cases of food poisoning are bacterial; in the UK the most common causes are *Campylobacter*, *Salmonella* spp., and *Clostridium perfringens*

Table 1.1 Different causes of food poisoning

Microbial infection/intoxication	– bacterial, e.g. salmonella/staphylococcus – viral, e.g. small round structured virus – fungal, e.g. aspergillus (aflatoxin) – protozoal, e.g. giardia
Microbial spoilage	– scrombroid poisoning from bacterial spoilage of fish such as mackerel and tuna, etc.
Poisonous animals	– shellfish poisoning caused by a 'red tide' of toxic dinoflagellates in plankton – ciguatera poisoning from fish living in tropical and subtropical waters following consumption of toxic dinoflagellates
Poisonous plants	– deadly nightshade, red kidney beans
Heavy metals	– copper, zinc, tin
Pesticides and herbicides	
Allergic or sensitivity reactions to certain foods	

Table 1.2 Laboratory-reported cases of food poisoning bacteria in England and Wales from 1984–1993

Year	Salmonella spp.*	Clostridium perfringens	Staph. aureus	Bacillus spp.	Total
1984	14 727	1 716	181	214	16 838
1985	13 330	1 466	118	81	14 995
1986	16 976	896	76	65	18 013
1987	20 532	1 266	178	137	22 113
1988	27 478	1 312	111	418	29 319
1989	29 998	901	104	164	31 167
1990	30 112	1 442	55	162	31 771
1991	27 693	733	61	95	28 582
1992	31 355	805	112	182	32 454
1993	30 652	669	28	41	31 390

*S. typhi, S. paratyphi and symptomless excretion of salmonella excluded. (Source: PHLS, London.)

Table 1.3 *Campylobacter* infections in England and
Wales from 1984–1989

Year	General outbreaks	All faecal isolates
1984	15	21 018
1985	17	23 572
1986	14	24 809
1987	17	27 310
1988	8	28 761
1989	9	32 526
1990	11	34 552
1991	8	32 636
1992*	10	38 552
1993*	5	39 385

(Source: PHLS, London.)
* Provisional

(Tables 1.2 and 1.3). Unlike salmonella the majority of campylobacter infections are sporadic cases from food sources (Table 1.3), and are rarely seen in large publicized outbreaks. However, it should be stressed that they are without doubt, the most frequently reported bacterial cause of diarrhoea in the UK.

When we try to understand the mechanisms of bacterial pathogenesis, it becomes apparent that food poisoning bacteria can be split into two main groups: those whose *modus operandi* includes little evidence of toxin production and those that cause disease primarily by the production of toxins. These two groups of bacteria are discussed in detail in Chapters 2 and 3 respectively.

Over the past few years, there have been on average at least 50 general community food poisoning outbreaks of unknown aetiology each year in England and Wales. It has been suggested that viruses may have played a role in some of these outbreaks. Viral food poisoning is still a very imprecise area; the available evidence is considered in Chapter 6. Chapter 6 also discusses protozoal species such *Giardia lamblia* and *Cryptosporidium parvum* whose pathogenicity and role in opportunist infections have been highlighted in recent years, primarily as a result of the AIDS epidemic. These organisms have been responsible for a small number of food poisoning outbreaks, and should not be overlooked when the more common microbial agents have been excluded.

Certain food products, particularly those of tropical origin, may become contaminated with fungi, which then multiply and produce mycotoxins, some of which can be fatal, even in small quantities. Mycotoxins may be implicated in a range of different diseases: recently, there has been considerable concern over the possible carcinogenic properties of aflatoxin associated with nuts and figs. Mycotoxic fungi as agents of food poisoning are dealt with in Chapter 5.

The causative agent of food poisoning may often be difficult to identify. Conclusive proof of a pathogen should require tests that comply with Koch's postulates (a set of conditions which should be fulfilled in order to establish that a given organism is the causal agent of a particular disease), but this is not always possible, and studies may be restricted to animal models and the detection of *in vitro* invasiveness of virulence-associated genes. It may be that some bacteria are facultative pathogens, their pathogenicity being dependent on such factors as the virulence of the strain, the infective dose and the susceptibility of the patient. Bacteria which have been commonly incriminated and may be included under this heading are members of the Enterobacteriaceae (e.g. *Klebsiella* spp., *Enterobacter* spp., *Citrobacter* spp.) and *Pseudomonas* spp. Because of their uncertain significance, we do not consider these organisms in any detail.

1.3 PATHOPHYSIOLOGY OF DIARRHOEA, VOMITING AND ABDOMINAL PAIN

Diarrhoea, vomiting and abdominal pain are probably the most common symptoms of food poisoning. While other clinical features such as fever may be seen with some regularity, these three symptoms are particularly important, and deserve further explanation here as a basis for understanding the mechanisms of pathogenesis of food poisoning organisms described in Chapters 2, 3 and 4. The spectrum and severity of clinical features associated with food poisoning may be useful guides as to which microbial agent is responsible, but they should not be thought of as inflexible; inconsistencies do occur. Reliable identification requires laboratory diagnosis, the main principles and techniques of which are outlined in Chapter 7.

1.3.1 Diarrhoea

Diarrhoea may be defined as the malabsorption of salt and water resulting in a net loss of fluid from the body (fluid secretion > fluid absorption), or the frequent evacuation of watery (or bloody) stools. It may occur when the intestinal mucosa is stimulated to secrete salt and water; when mucosal permeability is altered; following gross destruction of mucosal cells, or when there are disturbances of gastrointestinal motility and transit. Infective agents may produce diarrhoea by one or more of these mechanisms.

A number of bacterial toxins, e.g. cholera toxin, stimulate intestinal secretion and intestinal motility by an alteration in intracellular levels of secondary messengers, e.g. cAMP. These increased levels of cAMP alter the balance of Na^+ and Cl^- ions entering and leaving the enterocytes which directly affect water absorption and secretion. The net result is an efflux of Na^+ and Cl^- ions with more water being secreted than is absorbed (Figure 1.1).

1.3.2 Vomiting

Vomiting or emesis is the forceful expulsion of gastric and intestinal contents through the mouth. Vomiting is usually preceded by a feeling of nausea,

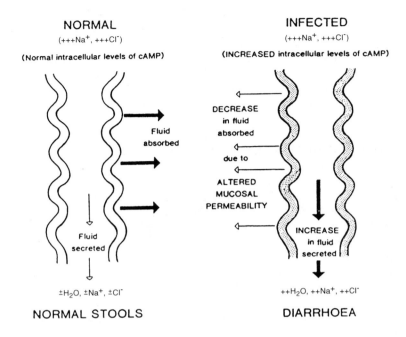

Figure 1.1   Diarrhoeal mechanisms associated with bacterial enterotoxins.

excessive salivation, pallor and sweating. In the vomiting process the stomach divides into two sections and a deep inspiration is taken; then with a strong contraction of the diaphragm and abdominal muscles the stomach contents are regurgitated up via the oesophagus.

The coordination of the different physiological events that occur during vomiting is under the control of the vomiting centre, which is situated in the fourth ventricle of the brain. The vomiting centre can be activated by afferent impulses arising from many parts of the body, especially the digestive tract. For example, the stomach and small intestine are innervated by afferent fibres in the vagus and splanchnic nerves. A large body of experimental evidence suggests that the vagus evokes vomiting with concomitant stimulation of the upper gastrointestinal tract. Stimulation from staphylococcal enterotoxin provides a good example of the implication of the abdominal vagus in emesis. Broadly speaking, abdominal vagal afferent fibres can be subdivided into those that originate from within the gastrointestinal mucosa (which is the point of contact with the food and any toxins it may contain), and those that constitute tension receptors within the muscular layers of the gut wall. These mechano-receptors signal the levels of tension generated and contractile activity in the gut.

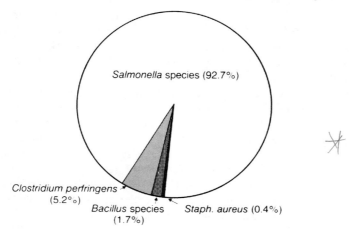

Figure 1.2   The four most common causes of bacterial food poisoning in England and Wales in 1988. (Source: PHLS, London.)

1.3.3 Abdominal pain

Unfortunately, little is known of the precise mechanism of abdominal pain production. The organs of the abdominal cavity have a lower number of nerve endings per unit of tissue compared with the skin; this contributes to the poor degree of localization of pain sensation in the abdomen. Moreover, it has been shown that inflammation in some way makes nerve endings more sensitive to mechanical and chemical stimuli. Any microbial infection which produces an inflammatory response may, therefore, make the host more susceptible to painful stimuli such as those which result from distension.

1.4 RECENT TRENDS IN FOOD POISONING IN THE UK

The 1980s witnessed a remarkable increase in the number of reported cases of bacterial food poisoning in the UK which has continued into the 1990s. During the 1970s, despite fluctuations from year to year, the figures showed only a modest upward trend; since 1980, however, there has been a sharp upturn in reports with a particularly steep rise since 1986 (Table 1.2). This rise is largely due to a dramatic increase in the number of cases of salmonella food poisoning, which in 1988 accounted for nearly 93% of the total (Figure 1.2) and has since remained relatively unchanged.

Increased reporting may be linked to laboratories becoming more aware of food poisoning and more skilled in diagnosing its causative agents; however, these factors alone could hardly explain the 65% rise in reported cases of salmonella between 1986 and 1988. There can be little doubt that this particular type of microbial food poisoning became very much more widespread in this country during the late 1980s and early 1990s. It should also be remembered that under-reporting is a significant problem where many types of food poisoning are concerned: a large number of outbreaks go unrecorded either because of the mild nature of the infection involved, or because it produces a high proportion of sporadic cases which result in no further epidemiological investigations. We will probably never know the full scale of food poisoning in this country. However, in 1994 a major food poisoning study was announced by the Department of Health in an attempt to find out the extent and cause of infectious intestinal disease in the community and the number of patients consulting their general practitioners. The objectives of the study were to estimate the true incidence of disease in the population and identify the associated microbiological agents, as well as determining possible factors that might be associated with a greater risk of becoming ill and estimating the economic costs of such illness.

The dramatic increase in reports of salmonella infections has been accompanied by a significant change in the relative isolation rates of the different serotypes (section 2.1.3) of this organism. The predominant serotype used to be *S. typhimurium*; this has now been replaced as the most frequently isolated serotype by *S. enteritidis*. Many *S. enteritidis* infections have been linked to consumption of uncooked eggs or egg products. The controversy in the late 1980s surrounding certain public statements about salmonella in eggs seems to have had the effect of raising public awareness of salmonella food poisoning, which may prove to be an important element in its control.

Other types of foodstuffs, such as soft cheeses and pâtés, were also the focus of public attention as possible sources of infection by *Listeria monocytogenes*. Although not commonly seen in the UK, listeriosis is a cause for concern, as it may lead to serious infections such as septicaemia and meningitis in infants, pregnant women, the elderly and the immunosuppressed. Fortunately more recent figures on the incidence of this organism as a cause of food poisoning have shown a decline. However, we need to remind ourselves of *L. monocytogenes* and other psychrotrophs in our increasing use of refrigeration and the popularity of chilled food which has an extended shelf life. A number of less common bacterial pathogens, including listeria, are considered further in Chapter 4.

L. monocytogenes is one of a group of organisms which have long been recognized as pathogens, but which have only been identified as causal agents of food poisoning in the last 10–15 years. Other bacteria which fall into this category are:

> *Campylobacter* spp.
> *Bacillus subtilis*
> *Bacillus licheniformis*
> *Aeromonas* spp.

Plesiomonas shigelloides
Yersinia enterocolitica and
Enteropathogenic *E. coli.*

In the same period, other, new bacteria have been discovered which are also implicated in food poisoning disease in the UK. These include:

Vibrio spp. (other than *V. parahaemolyticus*) and
Enterohaemorrhagic *E. coli.*

These two lists emphasize how dynamic this area of study has become in recent years. The epidemiology of food poisoning is considered in more detail in Chapter 8.

1.5 GLOBAL ASPECTS OF BACTERIAL DISEASE

Among an increase in emerging infections worldwide, a number of organisms such as salmonella, *E. coli* 0157 and *Cryptosporidium*, responsible for food poisoning, have recently caused significant outbreaks in the USA and have not surprisingly raised their public profile. It is too early to know what the concerted response will be to these infections, but it would probably be most beneficial if our understanding of these organisms could be enhanced, together with an overall tightening of public health measures. Of these organisms, salmonella is the most commonly reported cause of food poisoning not only in the UK but also in the USA and many other countries around the world. Since salmonella may be found in the intestinal tracts of humans and animals, it is easy to understand how widespread contamination of meats and vegetables can occur, and how difficult it can be to control infection. As in the UK, *S. enteritidis* has now replaced *S. typhimurium* as the predominant serotype in many parts of the world, although the reasons for this change are still unclear.

Campylobacter is known to be a common cause of food poisoning in developed countries. The same organism is also responsible for enteritis in children and adults, which may or may not be associated with food. In developing countries, on the other hand, campylobacter infections are very common in young children, but are generally followed by immunity. It is therefore unlikely that older children and adults will be as susceptible to campylobacter food poisoning as their counterparts in developed countries. In contrast to salmonella, which produces many general community outbreaks, almost all campylobacter food poisoning infections are sporadic cases – an epidemiological nightmare.

There is a very close relationship between the incidence of indigenous *Vibrio parahaemolyticus* infection and areas of warm coastal waters. Together with dietary practices such as eating raw fish, this fact explains why this organism is the most frequent cause of food poisoning in Japan, and why it is also common in certain parts of the USA, but is rarely encountered in the UK. A similar relationship between other newly discovered *Vibrio* spp. food poisoning cases and warm coastal waters has also been observed.

Food poisoning caused by *E. coli* (except serotype 0157) is fairly uncommon throughout the world, though the number of reported cases may not be a reliable indicator of its prevalence in certain areas. *E. coli* food poisoning is not usually seen in developed countries, because standards of sanitation and hygiene are normally high. However, this is often not the case in developing countries, where faecal contamination is more common. In these areas, laboratories might have difficulty recognizing and differentiating pathogenic strains from the many non-pathogenic strains of *E. coli* present in faeces.

However, in the UK since the late 1980s there has been a significant increase in the number of cases of *E. coli* 0157 food poisoning. Moreover, recently there have been notable large outbreaks in the USA which have caused a great deal of public concern.

Under-reporting may also need to be taken into account in the case of *Y. enterocolitica* food poisoning; this organism can take a long time to isolate and be identified and often fails to grow on standard enteric culture media at 37°C. Worldwide, this type of food poisoning is infrequently reported, but it is known that different serotypes are implicated in different countries. The picture is further complicated by the fact that some biotypes (strains distinguished by metabolic and/or physiological properties) and serotypes (strains distinguished antigenically) are less pathogenic than others. The importance of this organism lies in the fact that it can multiply at 4°C, the recommended temperature of most refrigerators.

Another organism which is capable of low temperature (4°C) multiplication is *L. monocytogenes*. Although listeriosis is not particularly common, there have been notable outbreaks in Western Europe and the USA. This organism is particularly important as it may cause septicaemia and/or meningitis and it has a high mortality rate of approximately 30%.

Problems arising from contamination with *Staph. aureus* will be seen worldwide wherever humans handle cooked foods, especially if those foods are incorrectly stored after handling. Since inadequate training or refrigeration facilities are often implicated *Staph. aureus* food poisoning is likely to be more widespread in developing than in developed countries. More general aspects of food hygiene are discussed in Chapter 11.

Cl. botulinum was thought to be less of a problem in most developed countries, although a number of recent incidents has cast some doubt on this. However, a great deal is known about its ecology and the conditions necessary for toxin production. Generally speaking, this knowledge allows us either to kill or inhibit the growth of the organism. Botulism is much more prevalent in those areas of the world, such as Alaska and China, where traditional fermented foods are still widely eaten, and where ignorance results in a lack of appropriate preventative measures.

Food poisoning caused by *Cl. perfringens* is usually associated with mass catering, notably the inadequate cooling of large quantities of contaminated cooked meat products. This disease is, therefore, frequently recorded throughout the world wherever such practices occur.

If all foodstuffs were consumed directly after cooking, there would be less potential for the development of toxins which may lead to food poisoning. Notoriously, *B. cereus* has been most often associated with cooked cereals and cereal products (especially rice) which have been stored incorrectly. *B. cereus* food poisoning will be found wherever cooked rice is improperly stored and inadequately reheated: Chinese take-away restaurants are a frequent source of infection in Western Europe.

1.6 FOOD SAFETY – CURRENT CONCERNS

Food safety was a very topical subject in 1989, when there was great public concern in the UK and elsewhere. National feelings about food safety culminated in recognition of the need for stricter controls and new legislation and led to the setting up of a Committee on the Microbiological Safety of Food chaired by Sir Mark Richmond; the Committee's findings have since become known as Richmond Reports I and II. The major recommendations of these reports are discussed in Chapter 10.

At the same time, other controversies arose concerning the possibility that humans could acquire BSE from consumption of contaminated beef and whether irradiated food is safe to eat. Only within the past few years has approval been granted for food irradiation in the UK, although it has been practised on a commercial scale, usually on spices, for a number of years in more than 20 countries (Table 1.4).

Irradiation involves passing rays from a radioactive or electron beam source through the food. It can eliminate or reduce the level of food poisoning bacteria. However, there are drawbacks: some bacterial spores and viruses are not affected at the doses used to kill vegetative cells; certain foods, such as those with a high fat content, do not lend themselves to this treatment, and the use of irradiation may lull producers and consumers into a false sense of security. It would probably make more sense to tackle the food contamination problem at source, thereby making irradiation unnecessary, and to continue to search for better alternatives when irradiation is advised. Although there is little, if any, proof, fears have been expressed over the possible harmful effects of radiation-induced free radicals in foods.

Over the past few years there has been a marked increase in the incidence of salmonella food poisoning and bacteraemia in patients with AIDS. As most salmonella infections can be avoided by eating well-cooked food and observing simple rules of food hygiene, food counselling is now being recommended. To decrease their risk of infection, all immunocompromised patients, including

Table 1.4 Countries presently irradiating food

Food	Country
Fruit	S. Africa
Apples	China
Garlic	Korea
Onions	Chile, Cuba, Germany, S. Africa, Thailand
Potatoes	China, Cuba, Japan, S. Africa
Spinach	Argentina
Dried vegetables	Netherlands, S. Africa
Grain	Commonwealth of Soviet Republics
Poultry	France, Netherlands
Fish	Netherlands
Shrimps and prawns	Netherlands
Frogs' legs	Netherlands
Sausages	Thailand
Cocoa powder or beans	Argentina
Spices	Argentina, Belgium, Brazil, Finland, France, Hungary, Iran, Netherlands, Norway, S. Africa, USA

those with AIDS, should now be counselled to cook raw food of animal origin, to avoid cross-contaminating cooked food with raw, and to avoid eating certain products, e.g. those containing raw eggs, which have been a common source of infection. Food counselling would also seem to be a useful approach to reducing the incidence of some of the serious infections which we have classed as food-related diseases (section 1.1).

Diet management has proved successful in the control of symptoms of patients suffering from irritable bowel syndrome (IBS). This condition, which may involve abdominal pain and diarrhoea, is caused by intolerance to specific foods or food components; it is in no way related to their microbial content. Diet management could also be used to reduce exposure to potentially danger-ous animal foods and their products (especially when raw), as they are responsible for the majority of food poisoning outbreaks. From this point of view it would appear beneficial to be vegetarian in eating habits; however, it should be pointed out that apart from obvious microbial food poisoning risks (e.g. listeria), other natural food-borne toxicants commonly occur in plants. Perhaps the most familiar natural toxicant is that associated with the toxicity of green potatoes containing high concentrations of glycoalkaloid. This may be responsible for neurological symptoms including apathy, restlessness, drowsi-ness and visual disturbances. A recent finding indicates that potato glycoalkaloids possess gut-permeabilizing activity which may point to the possibility of chronic effects of dietary glycoalkaloids; these will be enhanced in individuals who regularly eat large amounts of potato peel. Some other plants which also contain toxicants include cabbage, sprouts, courgettes (zucchini), beans and cassava.

Recently, many bacteria have been shown to be viable but non-culturable under conditions of stress and these include a number of human pathogens including *Campylobacter* spp., *Salmonella* spp. and *Shigella* spp. Under stress imposed by environmental extremes these organisms have been shown to enter a dormant state in which they cannot be detected by standard culture methods. The theory is that the presence of these 'dormant' organisms in food and water especially could pose a serious risk to public health if not detected. Although some of the questions surrounding the epidemiology of cholera can now be answered in the light of our understanding of viable but non-culturable organisms, it remains to be seen how important they are in the ecology of food poisoning bacteria.

One question which is not perhaps immediately obvious, and is rarely considered, is the economic cost of food poisoning. The main sources of economic losses are microbial spoilage between harvest or capture and con-sumption, and costs associated with food poisoning disease. More specifically, microbial spoilage is the result of the metabolic activity of organisms in a product which leads to adverse sensory changes. Over the years we have learned that each type of food is spoiled by a characteristic microflora or association and we have used this knowledge in attempts at preservation. The true costs of spoilage are poorly documented and rarely quantified, but they are considerable and are probably taken into account in the cost of the product. It is almost impossible to arrive at an accurate estimate of the cost of food poisoning, since we know that the majority of cases will go unreported, with the exception of major outbreaks. Even then, different analyses include different real and apparent costs in their breakdown. Expenses which would have to be taken into consideration include: medical care, including hospital costs; laboratory investigations; the cost to the economy generally or production and wages lost through illness; loss of production and market share for food manufacturers implicated in outbreaks, and legal expenses in the event of prosecution or claims for compensation. Furthermore, economic consequences to the company or establishment involved are often catastrophic. Figure 1.3 illustrates the size of the budget of the UK food chain, and shows how much money is tied up in food processing, manufacturing and distribution, in addition to the farm output. It is more than likely that the total annual losses worldwide due to food poisoning will exceed several thousand million pounds. This is a very realistic figure as in the USA alone for one particular organism (*E. coli* 0157) the estimated costs of medical treatment and lost productivity ranged from $200–500 million per year in 1993 and 1994.

These concerns all point to the need for a greater understanding of, and public commitment to, measures designed to control the spread of food poisoning organisms. The various aspects of microbiological control of food production, including a safety assurance system called Hazard Analysis: Critical Control Points (HACCP) are discussed more fully in Chapter 9. HACCP is a particular way of controlling food manufacture and handling, and covers

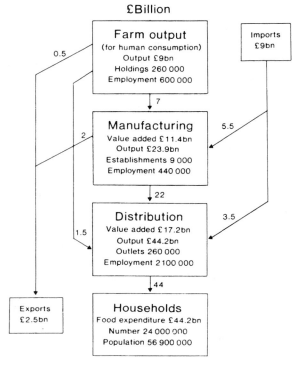

£Billion

Figure 1.3 U.K. food chain 1986. (Source: Waites, W.M. and Arbuthnott, J.P. (1990) *Lancet*, **336**, 722–5.)

product design, process design and operating practices. Hazard analysis involves the identification of factors which may have a dramatic effect on food safety. Once these factors have been identified, various critical control points can be introduced during processing to improve monitoring and so prevent or reduce risk to consumers. Recently in the USA the principles of HACCP have been reinforced and encouraged by the contribution of regulatory agencies such as the Food and Drug Administration, and industry. Together they were able to clarify the goal(s) of HACCP and how HACCP should be applied; provide uniformity in the application of HACCP principles during the development of HACCP plans; and assure that future guidelines would be consistent with the established principles of HACCP. HACCP is one of many significant new and encouraging developments outlined in this book which should help make the food we produce safer to eat.

2 Infective bacterial food poisoning

A.R. Eley

This chapter deals with a group of disease-causing bacteria, whose principal mode of pathogenesis is not thought to be toxin-mediated, even though most of them do produce some types of toxins. This group includes vibrio, yersinia and escherichia, as well as the two most common causes of bacterial food poisoning in the UK, salmonella and campylobacter (Table 2.1).

It used to be thought that in order to cause disease, these bacteria had to be present in large numbers (10^5–10^7 organisms/g food), as a result either of food being heavily contaminated initially, or of its having been stored in conditions which promoted the growth of large bacterial populations. However, it is now known that these organisms can cause disease with a relatively low infective dose (10^2 orgs/g), a fact which has obvious repercussions for food hygiene and laboratory diagnosis.

2.1 *SALMONELLA* SPECIES

These are Gram-negative, facultatively anaerobic, non-spore-forming bacilli that can be split into more than 2000 serotypes according to a system based on

Table 2.1 Basic characteristics of infective bacteria causing food poisoning

	Gram-reaction	Cell morphology	Catalase	Oxidase	Motility at 37°C	Growth at 4°C
Salmonella spp.	–	bacillus	+	–	v	v
Campylobacter spp.	–	bacillus	+	+	+	–
Vibrio parahaemolyticus	–	bacillus	+	+	+	–
Other *Vibrio* spp.	–	bacillus	+	v	+	–
Yersinia enterocolitica	–	bacillus	+	–	–	+
Yersinia pseudotuberculosis	–	bacillus	+	–	–	+
Escherichia coli (EPEC, EIEC)	–	bacillus	+	–	v	–

v, variable reaction.

somatic (O) and flagellar (H) antigens, known as the Kauffmann–White scheme (Table 2.2). Fortunately, human infection is limited to a small number of serotypes.

2.1.1 Pathogenesis

Following ingestion of salmonellae and passage through the stomach, the bacteria multiply and adhere to the brush border of the epithelial cells lining the terminal small intestine and the colon. The bacteria then penetrate the mucosal cells which results in damage, and then migrate to the lamina propria layer of the ileocaecal region. After further multiplication in the lymphoid follicles a leukocytic response develops, following reticuloendothelial hyperplasia and hypertrophy. This inflammatory response also mediates the release of prostaglandins, which stimulates cAMP and produces active fluid secretion which results in diarrhoea.

2.1.2 Clinical features and prognosis

Typically, the incubation period is approximately 12–36 hours. Usually the patient has a fever with abdominal pain and diarrhoea; vomiting is seen less frequently (Table 2.3).

Most patients with salmonella food poisoning usually recover within 7 days, and antibiotics are not usually recommended if only gastrointestinal symptoms occur. Rehydration may be required while the diarrhoea is acute and if it is particularly severe, and in debilitated patients, the very young and the elderly. During the diarrhoeal phase, personal hygiene should be scrupulous and food-handling should be avoided. In a small number of patients salmonella carriage and excretion may well last for several weeks, although more than 90% of patients are usually clear 10 weeks after the infection. Several severe complications of *S. enteritidis* have been reported and include a small number

Table 2.2 The Kauffmann–White scheme of serological classification: some examples

Serotype	O (somatic) antigen	H (flagellar) antigen	
		Phase 1	Phase 2
S. paratyphi A	1, 2, 12	a	–
S. paratyphi B	1, 4, (5), 12	b	1, 2
S. typhimurium	1, 4, (5), 12	i	1, 2
S. paratyphi C	6, 7, (Vi)	c	1, 5
S. typhi	9, 12, Vi	d	–
S. enteritidis	1, 9, 12	g, m	–

( ), antigens present only in some strains of the serotype.

Table 2.3 Clinical features of the illnesses produced by six major infective causes of bacterial food poisoning

	Salmonella	Campylobacter	V. parahaemolyticus	Y. enterocolitica	EPEC	EIEC
Incubation time (h)	12–36	72–120	12–24	24–36	12–72	12–24
Duration of illness (h)	≤168	72–120	48–120	<120	6–72	≤168
Vomiting	±	–	±	–	+	–
Nausea	+	+	+	–	+	–
Diarrhoea	+	+	+	+	+	+
Abdominal pain	+	+	+	+	+	+
Fever	+	+	±	+	+	+

–, rarely seen; ±, sometimes seen; +, often present.

of cases of acute renal failure, osteomyelitis and meningitis, all of which require appropriate antimicrobial therapy.

2.1.3 Incidence and epidemiology

Today, salmonella is the most common reported cause of food poisoning in the UK (approximately 90% of cases) and the USA (where numbers are increasing every year); it is also of importance in Japan and other areas around the world. Until 1988, the most common serotype in the UK was *S. typhimurium*. It has now been overtaken by *S. enteritidis*, which has been on the increase for a number of years, rose in 1988 to a level more than double that of *S. typhimurium* and has since increased its numbers proportionately (Table 2.4).

Approximately 80% of UK isolates of *S. enteritidis* belong to phage type 4 (PT4) (phage typing is a form of typing in which strains of bacteria are distinguished on the basis of differences in their susceptibilities to a range of bacteriophages). A similar trend of increased incidence of *S. enteritidis* food poisoning has also been observed in the USA and in other European countries. *S. typhimurium* however, is still important particularly in the UK, where one type, DT104 has increased considerably since 1990. This increase has been largely due to to a strain which is resistant to ampicillin, chloramphenicol, streptomycin, sulphonamides and tetracyclines.

Classically, *Salmonella* species such as *S. typhi* and *S. paratyphi* are responsible for outbreaks of typhoid and paratyphoid, so-called enteric fevers, which are rarely food-borne and not usually regarded as food poisoning (Chapter 1). Occasionally, other *Salmonella* species may cause urinary tract infections and bacteraemia which may lead to meningitis and osteomyelitis. These are seen much more frequently in immunocompromised patients, especially those with AIDS, producing a 20-fold increase in gastroenteritis and bacteraemia as compared with normal patients.

Table 2.4 Laboratory reports from England and Wales of salmonella isolations to the Communicable Disease Surveillance Centre

Year	S. typhimurium	S. enteritidis
1984	7264	2071
1985	5478	3095
1986	7094	4771
1987	7660	6858
1988	6444	15 427
1989	7306	15 773
1990	5451	18 840
1991	5331	17 460
1992	5401	20 094
1993	4779	20 254

(Source: PHLS, London.)

For epidemiological tracing in outbreak management where strains share a common phage type, plasmid analysis (section 7.3.6) may be useful in further subdividing them.

2.1.4 Ecology

Salmonella organisms are widely distributed in nature (Figure 2.1) and are commonly found in the intestinal tracts of animals and human carriers.

They are excreted in faeces and in this way may contaminate any objects with which faeces come into contact. Salmonella are distinguished from other food poisoning organisms by certain important features which include:

1. frequent presence in certain common foods, e.g. in 1987 60% of super-market chickens were found to be contaminated;
2. the ability to grow in a variety of foodstuffs over a wide temperature range, multiplying to give large populations in some cases;
3. ease of dissemination and spread from person to person;
4. a prolonged period of excretion after recovery (including a carrier state).

Salmonellosis is the classical form of microbial food poisoning in which human beings may be implicated as carriers, posing a further potential source of infections and outbreaks. It has been estimated that as many as 50 000 people may be excreting salmonellae at any one time in the UK.

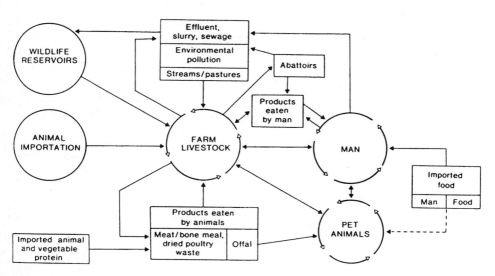

Figure 2.1 Cycle of salmonella transmission. (Source: Baird-Parker, A.C. (1990) *Lancet*, **336**, 1231–5.)

2.1.5 Foodstuffs

A number of foodstuffs have been associated with salmonellosis, typically meat products (especially poultry), eggs, untreated milk and dairy products, and any foods that have undergone faecal contamination (Table 2.5). The importance of milk pasteurization is emphasized by the fact that there have been no outbreaks of salmonellosis due to milk in Scotland since the introduction in 1983 of compulsory heat treatment of all milk for sale to the public.

During the past few years England and Wales have experienced an epidemic of food poisoning caused by *S. enteritidis* PT4. Epidemiological, veterinary and microbiological findings indicated that this salmonella was coming from contaminated carcasses, from eggs which had become infected through damage after laying, and also from a new source namely the contents of intact hens' eggs. The proportion of eggs that are infected before laying is highly variable (ranging from <0.1 to 20%, although for most it is very low), but because many millions of eggs are consumed daily the number of human infections originating from such contaminated eggs represents an important public health problem on a national scale.

In the USA a significant increase in human infections with *S. enteritidis* was observed in the North-Eastern states towards the end of 1986. Many of these outbreaks were associated with raw or lightly cooked shell eggs. Similar outbreaks were also recognized in some mid-Atlantic and south-Atlantic states in 1988. Overall in the USA *S. enteritidis* caused 375 recognized outbreaks between 1985 and 1991. As in the UK, veterinary investigations indicated that the contents of intact eggs were infected as a result of trans-ovarian infection resulting in vertical transmission. Interestingly, *S. enteritidis* PT4 was not detected and *S. enteritidis* PT8 and PT13A were instead found to be the most common.

S. enteritidis PT4 strains have also been isolated from human cases in other Western European countries and Scandinavia. It is worth remembering that in 1987 30% of *S. enteritidis* PT4 infections were found in travellers recently returned from abroad, most notably from Iberia and adjoining islands. In Spain,

Table 2.5 Foodstuffs commonly associated with bacterial causes of infective food poisoning

Bacteria	Meat and/or meat products	Poultry	Eggs	Milk	Dairy products	Seafood and/or shellfish	Vegetables	Cereals
Salmonella	✔	✔	✔	–	✔	–	–	–
Campylobacter	✔	✔	–	✔*	–	–	–	–
Vibrio spp.	–	–	–	–	–	✔	–	–
Y. enterocolitica	✔	–	–	✔	✔	–	–	–
EPEC/EIEC	✔	–	–	–	–	–	✔	–

*Unpasteurized milk

home-made mayonnaise has emerged as an important contributor to this problem. In the Basque region *S. enteritidis* was found to be responsible for 78% of outbreaks of known aetiology, with eggs and egg-based products accounting for 90% of outbreaks caused by salmonella.

2.1.6 Control

The control of salmonella in the food chain is a complex matter, because of the interrelationship between environmental contamination, farm animals and humans (Figure 2.1). The increasing trends in human infections, and recent food-borne outbreaks involving *S. enteritidis* in eggs emphasize the need for increased vigilance and concerted government and industrial controls in all aspects of food production.

There are a number of ways of controlling the entry and spread of salmonella in animals. These include regulated importation of live and slaughtered animals, use of salmonella-free breeding stocks and animal feeds, and good husbandry practices with egg-laying and broiler flocks. More recently a number of workers have concluded that salmonella infection can be significantly reduced by vaccination of poultry flocks.

The education of consumers and of food-handlers on the safe handling and cooking of potentially hazardous meats and other raw ingredients is of paramount importance. Knowledge of basic food preparation techniques, such as adequate refrigeration and thorough cooking of foods, should be stressed at all levels. Salmonella infections commonly arise after the consumption of certain raw or uncooked foods in which the bacteria are often inherently present. Adequate cooking, however, usually eliminates the risk of infection as neither *S. enteritidis* nor *S. typhimurium* is heat resistant. In the case of contaminated eggs, some traditional methods of cooking, such as poaching and soft boiling, have proved to be inadequate. Recent tests have shown that any form of cooking where all or some of the yolk remains liquid can permit the survival of *S. enteritidis* PT4, even from a relatively small inoculum. These results add further weight to the guidelines issued by the Department of Health in England and Wales to hospital caterers, advising that raw or lightly cooked eggs should not be served to hospital patients. It must be stressed that appropriate methods of cooking eggs and other foodstuffs must be accompanied by sound kitchen hygiene, to prevent cross-contamination of raw and cooked foods, which can be another significant cause of infection.

2.2 *CAMPYLOBACTER* SPECIES

These are Gram-negative microaerophilic bacilli. 'Microaerophilic' means that these organisms prefer to grow in an environment of approximately 5% O_2; this unusual growth requirement explains why they were not isolated routinely in diagnostic laboratories until some years ago. A further characteristic feature of these organisms is that their optimal temperature for growth is 42°C. These bacterial cells appear 'vibrio-like' as they are slender, spirally curved rods, non-spore-forming, and are oxidase-positive. Although these organisms were discovered many years ago, it was not until a selective medium was used in 1977 to improve their isolation that their importance as a cause of gastroenteritis was recognized. *C. jejuni* is the most commonly isolated species in the UK although *C. coli* may also be a significant pathogen in certain countries, e.g. the former Yugoslavia.

2.2.1 Pathogenesis

The mechanism of pathogenesis, including the induction of diarrhoea, is uncertain. Several potentially pathogenic properties have been identified: attachment and colonization, invasiveness, and cholera-like enterotoxin and cytotoxin production. However, many studies have reported conflicting results, which suggests that some pathogenic mechanisms may still not have been identified, and that pathogenesis is almost certainly multifactorial.

2.2.2 Clinical features

Following an incubation period of approximately 3–5 days there is often severe abdominal pain, fever, and bloody diarrhoea with nausea but with little or no vomiting. As a result of intestinal tissue injury, mucus, blood and faecal leucocytes can often be seen in the stool under microscopy.

2.2.3 Prognosis

The illness may well persist for 3–5 days and usually clears up within 1 week. Rarely, symptoms persist for longer than a week, and if this situation arises, the antibiotic ciprofloxacin (which is now replacing erythromycin in the UK) may be prescribed, usually resulting in recovery within a few days. In a small number of patients with other underlying conditions, complications such as septicaemia, meningitis, cholecystitis and respiratory infections may arise.

2.2.4 Incidence and epidemiology

Campylobacter spp. one of the most common causes of diarrhoea in many countries, have been implicated as a cause of travellers' diarrhoea, and are considered to be a very frequent cause of food poisoning in the UK and in the USA (Table 1.3). It is difficult to put a reliable figure on the number of cases of campylobacter food poisoning in the UK, as instead they are recorded as the total number of faecal isolates from infected patients (whose contaminated source might not always be from food). Another problem is that these organisms are usually responsible for a large number of sporadic cases where the source is unknown; few outbreaks are recorded.

In the context of diarrhoeal disease, campylobacter enteritis is commonly seen as a symptomatic infection in developed countries (UK, USA, Western Europe, Scandinavia, Australia, Canada), whereas in developing countries (Bangladesh, Peru, Rwanda and the Gambia) it is hyperendemic and the rate of asymptomatic infection is very high. In the developing countries, isolation rates are highest during the first 2 years of life, and then decline rapidly as immunity develops. Host factors and strain differences may offer some explanation of such variation in the natural history of campylobacter infections in different parts of the world. It would seem reasonable to suggest that *Campylobacter* spp. are probably less important as a cause of food poisoning in developing countries than they are in developed countries.

In the past, most outbreaks of campylobacter food poisoning have been traced to unpasteurized milk. Because these organisms are biochemically almost inactive, the usual method of tracing outbreaks involves one of two serotyping schemes (Lior and Penner) to differentiate between strains. It should also be remembered that different serotypes predominate in different countries.

2.2.5 Ecology and foodstuffs

Campylobacter spp. are often found in the intestinal tracts of many types of animals, including domestic dogs and cats, and especially birds and chickens. There is now general agreement that the environment is the major source of campylobacter in farm animals, and studies on poultry have confirmed this. For the prevention of colonization of chickens, hygiene barriers in the broiler house could be a cost-effective control measure.

Since these organisms are often present in the intestinal tract of animals, sources of infection include under-cooked chicken and meat, unpasteurized milk and water contaminated with animal or bird faeces. Although *Campylobacter* spp. are not particularly hardy in food or in the environment and are rapidly killed during cooking, only a low infective dose (reported by some to be as low as 2×10^2 orgs/g) is necessary to produce disease, as all multiplication occurs in the gastrointestinal tract. Cross-contamination of cooked food with raw poultry in the kitchen is, therefore, one likely means of disease transmission, as the few organisms involved are sufficient to cause infection.

2.2.6 Control

Fortunately *Campylobacter* spp. are fastidious in their growth requirements and are susceptible to heat so that they are usually killed by the majority of cooking procedures used to kill enteric bacteria in foods. They are nonetheless responsible for high numbers of gastroenteritis cases, in which cross-contamination of raw and cooked food products is thought to play a major part. As these organisms have such a low infective dose, even the smallest amount of contaminated raw food left in the food preparation area could be sufficient to cause infection.

The most important methods of protecting the public are through improved education and reinforcement of hygienic practices at the strictest level. Emphasis should be placed on segregating raw products from any food or material that will directly or indirectly come into contact with the consumer.

2.3 VIBRIO PARAHAEMOLYTICUS

These are Gram-negative, facultatively anaerobic, non-spore-forming bacilli which are oxidase-positive and halophilic, i.e. they require salt for optimal growth.

2.3.1 Pathogenesis

The mechanisms of pathogenesis are not fully understood at present, although some evidence suggests a direct invasion of the mucosa. It is also known that some strains produce a range of bacteriologically active toxins, and that almost all pathogenic strains produce a thermostable direct haemolysin (TDH). This causes beta-haemolysis of human erythrocytes in a blood agar medium (total breakdown of red cells producing a clearing) and is responsible for the reaction known as the 'Kanagawa phenomenon'. The association between the ability of an isolate to produce TDH and its ability to cause gastroenteritis is well established. However, there has been at least one outbreak of gastroenteritis caused by Kanagawa-negative strains, which suggests that TDH is not the only virulence factor of V. parahaemolyticus.

2.3.2 Clinical features and prognosis

The incubation period may be from 4 to 48 hours, but is usually 12 to 24 hours after ingestion of contaminated food. Characteristic symptoms are a profuse diarrhoea, abdominal pain and nausea, sometimes with fever and vomiting.

The illness is usually self-limiting within 2–5 days, and no specific treatment is necessary unless severe fluid loss has occurred.

2.3.3 Incidence and epidemiology

This organism was first described in 1951 and is the most frequent cause of food poisoning in Japan. It is also a cause of travellers' diarrhoea and locally acquired wound infections when present in large numbers in the coastal marine environment.

Although rarely encountered in the UK (only five cases between 1984 and 1988), *V. parahaemolyticus* is more prevalent in the USA and in other countries with warm coastal waters. The organism is discouraged by low water temperatures (< 10°C), and this explains its absence from British coastal waters except during the summer months. In the UK it is usually isolated from patients who have eaten food such as frozen seafoods imported from parts of Asia.

For epidemiological tracing, serotypes, biochemical characteristics and plasmid profiles may be compared.

2.3.4 Ecology

Since human pathogenic vibrios occur naturally in aquatic environments, the microbial ecology of these organisms is important in understanding the occurrence and epidemiology of human infections. Apart from water temperature, four other environmental factors are important:

1. the concentration of organic material in the water;
2. salinity;
3. the potential for association with higher organisms (e.g. plankton, shellfish and fish);
4. association with sediments, which are thought to harbour the organisms in a dormant state during the cooler winter months.

2.3.5 Foodstuffs

V. parahaemolyticus food poisoning is almost exclusively associated with the consumption of raw or lightly cooked contaminated seafood (shrimps, prawns, crab, lobster, fish) or shellfish (oysters, clams). There is a pronounced seasonal incidence, with outbreaks occurring mostly during the summer months, when *V. parahaemolyticus* is most prevalent in the aquatic environments from which seafood and shellfish are harvested. Unfortunately, this organism may be actively concentrated in the gut of filter-feeding shellfish and may proliferate in contaminated seafoods stored without refrigeration. There is a significant risk from raw seafood dishes prepared and eaten in Japanese-style restaurants, especially if the ingredients have been improperly stored, allowing levels of *V. parahaemolyticus* to rise above the infective dose of 10^6 orgs/g of food.

2.3.6 Control

As this organism is ubiquitous in estuarine waters of temperate seas, any seafood harvested from these waters, especially in warmer months, should be regarded as contaminated. In addition, the ability of *V. parahaemolyticus* to grow rapidly at low temperatures will allow initial low bacterial counts to reach dangerous levels under improper conditions of storage. Where the consumption of raw seafood is concerned, there will always be a danger of ingesting sufficient numbers of a pathogenic strain to cause gastroenteritis.

The proper cooking of seafood is the only method currently available to inactivate *V. parahaemolyticus*, though it will not affect any preformed thermo-stable Kanagawa haemolysin which would remain stable in the cooked food. Cooking is obviously not effective, however, if cross-contamination with raw product is allowed to take place following the cooking process; strict hygiene measures are also essential.

2.4 OTHER *VIBRIO* SPECIES

Evidence over the last few years from epidemiological and ecological studies has established that there are many *Vibrio* species other than *V. parahaemolyticus* (and *V. cholerae*-01 serotype) which are capable of causing a wide range of diarrhoeal and systemic diseases in humans. Perhaps the most important species to date are *V. cholerae* non-01 serotypes, *V. vulnificus*, *V. fluvialis*, *V. mimicus* and *V. hollisae* (Table 2.6).

It is worth stressing that these organisms are quite readily killed (with the possible exception of *V. hollisae*) by normal cooking procedures, and that problems generally arise after consumption of raw or lightly cooked seafood. As most of the research and reports have come from the USA, where cases have been recorded in relatively small numbers, it is difficult to form a global picture of the importance of these organisms in causing food poisoning. A further two vibrios, *V. furnissii* and *V. metschnikovii*, have been isolated from a very small number of cases of food poisoning, usually after consumption of seafood.

2.4.1 *V. cholerae* non-01 serotypes

The term food poisoning does not normally encompass the enteric fevers typhoid or cholera, but some mention of the latter needs to be made in the context of *V. cholerae* non-01 serotypes.

Epidemiological and ecological studies have strongly implicated the natural aquatic environment as a source of *V. cholerae* 01 in sporadic and epidemic cholera outbreaks. This is still true in non-endemic regions where living standards are generally high, such as the USA, Australia and Italy. However, counts of free-living *V. cholerae* 01 in freshwater and saline environments are usually much lower than the necessary minimum infective dose. It has been suggested that in these environments the concentration of organisms by fish

Table 2.6 Characteristics of *Vibrio* species

Property	V. parahaemolyticus	V. cholerae	V. vulnificus	V. fluvialis	V. mimicus	V. hollisae
Oxidase	+	+	+	+	+	+
Growth in:						
0% NaCl	−	+	−	−	+	−
6% NaCl	+	v	+	+	v	v
Possession of:						
Lysine decarboxylase	+	+	+	−	+	−
Arginine dihydrolase	−	−	−	+	−	−
Ornithine decarboxylase	+	+	+	−	+	−
Acid production from:						
Lactose	−	−	v	−	v	−
Sucrose	−	+	−	+	−	−

−, negative reaction; v, variable reaction; +, positive reaction.

and shellfish may represent the first stages in the increase in numbers required to achieve an infective dose. However, fish and shellfish are themselves only rarely implicated in the aetiology of cholera, which is still predominantly caused by faecal contamination of water.

Since *V. cholerae* non-01 serotypes (previously called 'NAG' – non-agglutinable – or 'NCV' – non-cholera – vibrios) occur more frequently and in higher numbers than *V. cholerae* 01, similar bioconcentration of these organisms in shellfish may be a more significant factor in causing disease in humans. This may explain the association between this type of food poisoning and the recent consumption of seafood or shellfish. Although our knowledge is incomplete, there are also indications that *V. cholerae* non-01 serotypes may be more capable than *V. cholerae* 01 of surviving and multiplying in a wide range of other foods.

Patients who have been infected with *V. cholerae* non-01 serotypes typically produce symptoms after travelling abroad or after ingesting seafood or shellfish. Although some symptoms may be so severe as to mimic cholera, typical features include diarrhoea (which can be bloody), abdominal cramps and occasional vomiting and fever.

V. cholerae non-01 serotypes produce a broad spectrum of virulence factors including cholera-like toxin, enterotoxin, cytotoxins and haemolysins, although their precise role in the pathogenesis of these diseases is generally unknown.

It remains to be seen whether *V. cholerae* non-01 serotypes will be considered more important food poisoning bacteria on a worldwide scale in future years.

2.4.2 *V. vulnificus*

Although the pathogenic mechanisms of this bacterium are not yet fully understood, they may be related to the presence of a capsule, which has been correlated with virulence; to the production of extracellular enzymes which may contribute to host tissue damage; to the presence of outer membrane proteins which may serve as siderophores (iron-chelating compounds), to the presence of pilus-like structures and the ability to adhere to human epithelial cell lines, and to the presence of toxins.

This halophilic vibrio is now associated with three distinct clinical syndromes:

1. Primary septicaemia characterized by a mortality rate of 50% with a rapid onset (< 24 hours), following ingestion of contaminated raw bivalve shellfish. This condition is often accompanied by fever, abdominal pain, vomiting and diarrhoea. It is thought that the organism enters the bloodstream via the portal vein or the intestinal lymphatic system. It has been reported that patients with immunodeficiency, liver disease or abnormalities of iron metabolism may exhibit a rapid, often fatal sepsis.

2. Wound and soft-tissue infections, with a mortality rate of approximately 20% (usually within 24 hours of the onset of symptoms). These infections usually follow contamination of a wound sustained during activities associated with saline aquatic environments. Although fever and vomiting are commonly seen there is little diarrhoea and no abdominal pain.
3. More rarely, acute diarrhoea after consumption of raw oysters. There is no accompanying septicaemia and although often prolonged the diarrhoea is self-limiting.

Although *V. vulnificus* infection is rarely encountered, the organism is of particular concern as it is widely and sporadically distributed in US coastal waters, and normal indicators of water and shellfish quality do not ensure its absence. As with *V. parahaemolyticus*, there is a striking trend towards a greater frequency of infection during warmer months of the year, when *V. vulnificus* is most abundant in the aquatic environment. Infections are more likely to occur in areas where water temperature remains high for most of the year; this explains the predominance of reported disease in the mid-Atlantic and Gulf coast states of North America.

2.4.3 *V. fluvialis*

This organism has emerged as a potential enteropathogen since it was first reported in 1977. Although the mechanisms of pathogenesis are unknown, it is thought that enterotoxins have a role to play in causing gastroenteritis. There have, however, been few reports of food poisoning illness caused by this organism; most have been from the USA and were associated with raw or under-cooked seafood, especially oysters.

    V. fluvialis has been isolated from Louisiana coastal waters, the Adriatic sea, Hong Kong and from seafood and shellfish present in surrounding areas. As with certain other *Vibrio* species, the isolation of this organism is rare in aquatic environments in cooler climates, but it is found in larger numbers in warmer waters.

2.4.4 *V. mimicus*

This is a relatively newly described pathogen and its virulence is unclear, although the presence of an enterotoxin and a toxin almost identical to cholera toxin have been reported. Gastroenteritis is a common feature, frequently after consumption of seafood and shellfish, particularly raw oysters. Reports of sporadic incidents have always been linked with a close proximity to warm seawater: countries involved include Canada, Mexico, Bangladesh, the Philippines and New Zealand.

2.4.5 V. hollisae

This organism is another newly described pathogen; it contains a haemolysin similar to TDH of *V. parahaemolyticus* and an enterotoxin, although the importance of these as virulence factors is still unclear.

Despite a lack of information about its ecology and capacity for survival in marine environments, this organism has occasionally been associated with septicaemia and cases of diarrhoea in and around the USA, especially in warm seawater areas such as the Gulf of Mexico (although some cases have been reported in Maryland, which is close to cooler waters). There is a strong association between *V. hollisae* infections and consumption of raw seafood or fish which has been fried or treated by drying and salting. The latter indicates a potential for this vibrio to survive some methods of cooking and preservation.

Of diagnostic importance is the fact that some strains fail to grow on thiosulphate citrate bile salts (TCBS) agar (a common selective medium of alkaline pH for *Vibrio* spp.) and may not therefore be isolated on routine examination.

2.5 YERSINIA ENTEROCOLITICA

These are Gram-negative, facultatively anaerobic, non-spore-forming bacilli or coccobacilli which are psychrotrophic, i.e. they can multiply at refrigeration temperatures (4°C). This organism is a member of the family Enterobacteriaceae and is related to salmonella, shigella and escherichia.

2.5.1 Pathogenesis

Food poisoning is caused by the consumption of food containing viable pathogenic *Y. enterocolitica*; the infective dose is, however, unknown. Mechanisms of pathogenesis are unclear, but tissue invasion is thought to be an important factor. Although an enterotoxin has been found, it has been reported that production does not take place above 25°C. Other possible virulence factors include the production of fimbriae; the presence of other membrane proteins which are protective against host defences; the possession of a plasmid which is not observed in environmental isolates or non-pathogenic strains, and the capacity for intracellular survival.

2.5.2 Clinical features and prognosis

The incubation period is typically 24–36 hours after ingestion of contaminated food, but may extend to between 3–5 days. Often there is abdominal pain which may be so severe as to resemble acute appendicitis; this is usually accompanied by fever and diarrhoea, but rarely by nausea and vomiting. As diarrhoea is not always present, yersiniosis has been responsible for many unnecessary appendicectomies. In general, this organism usually causes intestinal syndromes of varying severities (typically an enteritis), and extraintestinal infections which may develop into septicaemia, especially in immunocompromised patients. In addition, infection may also lead to several autoimmune conditions, notably erythema nodosum, ankylosing spondylitis, Reiter's syndrome and polyarthritis.

The duration of the illness is usually a few days, and it is followed by complete recovery, except in the case of those patients who develop the autoimmune diseases mentioned above.

2.5.3 Incidence and epidemiology

The incidence of Y. enterocolitica in food poisoning may well be under-reported, since it can take a long time to isolate and identify the organism, and not all laboratories are familiar with it. Moreover, it often fails to grow on standard enteric media at 37°C, but will do so at 25°C. In this context, Y. enterocolitica has been isolated only rarely in the UK, USA and Japan, with a small number of outbreaks and sporadic incidents reported. Although this organism plays a more important role in causing enteritis in other parts of Western Europe and Scandinavia, there does not seem to be a concordant pattern of higher incidence of food poisoning in these countries. A further complication lies in the fact that certain biotypes and serotypes of Y. enterocolitica are less pathogenic than others, and that even among the so-called pathogenic serotypes (0:3, 0:8, 0:9) there are non-pathogenic strains. Moreover, there is still controversy over the pathogenicity of other serotypes.

Epidemiological studies have shown differences in isolation rates and in the distribution of pathogenic serotypes in different parts of the world. Serotype 0:3, which predominates in Europe, is rarely encountered in the USA as a cause of food poisoning. Similarly, serotype 0:9 has been responsible for outbreaks in Europe and Japan although it is rarely seen in the USA. Food poisoning outbreaks which have been reported in the USA traditionally involved serotype 0:8, typically isolated from milk. In recent years there has been a shift towards serotype 0:3 as the principal cause of outbreaks associated with Y. enterocolitica in the USA.

2.5.4 Ecology

This organism is not uncommon in the environment or in food and was found to be present in 3.5% of faecal specimens submitted to a Public Health Laboratory in England over a 12-month period in the mid-1980s. Y. enterocolitica has been isolated from various animals including pigs, dogs, cats, monkeys and rats, which could suggest an animal reservoir of infection.

2.5.5 Foodstuffs

Y. enterocolitica has been particularly associated with milk and milk products and was isolated from almost 50% of raw milk samples analysed in a recent study in the UK. Controversy has arisen over its resistance to heat, as outbreaks have been traced to pasteurized milk, although in theory the organism should not survive pasteurization. There is cause for concern in the fact that *Y. enterocolitica* has been isolated from high-temperature, short-treatment (HTST) milk: it should be remembered that the organism is capable of multiplying even in cold storage. Although isolation has also been reported from vegetables and many different types of meats and meat products, it is not known whether the strains concerned were serotyped and deemed to be pathogenic.

2.5.6 Control

From an epidemiological standpoint it appears that there may be a link between carriage of *Y. enterocolitica* in pigs and an increase in human infection. It has been suggested that pork products and raw meat mixtures containing contaminated pork may be important in transmission. (There is still some speculation about this, because of our inability to isolate small numbers of *Y. enterocolitica* from large populations of related bacteria.) To attempt to control *Y. enterocolitica* food poisoning by eliminating the organism from the animal host would be a monumental task, on at least the same scale as the control of salmonella in animals.

In the USA some food poisoning outbreaks caused by *Y. enterocolitica* have indicated that human carriers may be important in transmission. This underlines the need for education of food-handlers, who should be encouraged to behave responsibly in their work both before and following gastrointestinal symptoms.

However, until more is discovered about the role of this organism in food poisoning we can offer little practical help in controlling it. Despite the controversy over heat resistance, it would seem that *thorough* cooking is likely to prove effective, provided that proper standards of hygiene are observed in the handling and storage of cooked dishes. As the organism is a psychrotroph, with the ability to grow at refrigeration temperatures, cross-contamination to other foods is a particular problem.

2.5.7 *Y. pseudotuberculosis*

This organism, which is closely related to *Y. enterocolitica*, may have been responsible for some sporadic cases and outbreaks of food poisoning in Europe and Japan. Clinical signs were similar to *Y. enterocolitica*, with abdominal pain being the major or the only symptom of disease. However, although foodstuffs were suspected of harbouring the organism, an element of doubt remained as to the causation of disease.

2.6 *ESCHERICHIA COLI*

These are Gram-negative, facultatively anaerobic, non-spore-forming bacilli which have been subdivided into a number of groups. The four most important groups are:

1. Enteropathogenic   (EPEC) ⎱ where the pathogenic mechanisms
2. Enteroinvasive   (EIEC) ⎰ are still under investigation

3. Enterotoxigenic   (ETEC) ⎱ where pathogenicity is related to
4. Enterohaemorrhagic (EHEC) ⎰ toxin production

The last two groups of *E. coli* will not be considered further here – they are discussed in detail in Chapter 3.

2.6.1 Pathogenesis

Although *E. coli* is part of the normal flora of the large intestine of humans and animals, most strains at this site are non-pathogenic. However, some strains can produce enteric, urinary tract and wound infections as well as food poisoning, and occasionally septicaemia and meningitis. These organisms are divided on the basis of serotyping of the O antigens in a similar manner to salmonella. Although there are hundreds of serotypes of *E. coli*, only a relatively small number commonly cause infection (Table 2.7).

In general, evidence suggests that *E. coli* can multiply in food and that large numbers of organisms (e.g. 10^5–10^7 orgs/g) need to be present to cause infection. Some progress has been made in defining modes of pathogenesis, though it must be stressed that the exact mechanisms are still unknown.

Where EPEC strains are concerned, the suggestion is that some form of enterotoxic product is elaborated, which differs from the heat-labile toxin (LT) or heat-stable toxin (ST) found in ETEC isolates, although the evidence for this is often contradictory. EPEC strains have been examined for toxins that can induce

Table 2.7 Major O serotypes of EPEC and EIEC associated with diarrhoeal disease

---

Enteropathogenic
   018, 044, 055, 086, 0111, 0114, 0119, 0126, 0127, 0128ab, 0142, 0158

Enteroinvasive
   028ac, 029, 0124, 0136, 0143, 0144, 0152, 0164, 0167

---

fluid secretion from the intestine, and also for cytotoxins. Again, the evidence is confusing, but some EPEC strains have been shown to produce Shiga-like toxin (SLT) (the toxin found in *Shigella* spp. which is responsible for bacillary dysentery), albeit in low concentrations. The other major virulence factor proposed is adherence: the intimate attachment of EPEC strains to intestinal mucosa could disturb the function of the microvilli and so cause diarrhoea.

EIEC serotypes (Table 2.8) closely resemble shigella in many respects (section 4.2). Like shigella, their major pathogenic feature is the capacity to invade and proliferate within epithelial cells and to cause eventual death of the cell. This invasive capacity is related to the presence of a large plasmid which codes for the presence of several outer-membrane proteins associated with invasiveness.

2.6.2 Clinical features and prognosis

After an incubation period of 12–72 hours, symptoms consist mainly of diarrhoea, sometimes with the presence of blood and/or mucus. In general, symptoms may vary in type and severity depending on the serotype of the organism.

After a few days, the infection is usually self-limiting and with the exception of severe diarrhoea, in which patients may require rehydration, no other form of treatment is usually necessary.

2.6.3 Incidence and epidemiology

Food poisoning caused by EPEC and EIEC is apparently fairly uncommon in both the UK and the USA; there are few reports of outbreaks worldwide. A

Table 2.8 Similarities in biochemical characteristics of *E. coli* and *Shigella* spp.

Media or test	E. coli	Shigella *spp.*
Lactose fermentation	+	– or slow
Dulcitol fermentation	v	–
Gas from glucose	+	–
Malonate	–	–
Citrate	–	–
Urease	–	–
Methyl red	+	+
Lysine decarboxylase	v	–
Arginine decarboxylase	v	v
Ornithine decarboxylase	v	v
H_2S	–	–
Gelatin hydrolysis	–	–
Indole	+	v

v, variable reaction.

major problem exists in relation to these organisms, however, in that they are difficult to recognize and differentiate from many other non-pathogenic strains of *E. coli* on routine culture media. Usually, serotyping is required to identify strains with any degree of certainty.

2.6.4 Ecology and foodstuffs

Faecal contamination of foods, either by direct contact or indirectly via component water, is probably the most important method of transmission. This type of contamination is most likely to affect meats, meat products and fresh vegetables, which then become the source of infection. Since *E. coli* are always present in faeces, they are often used as a marker of faecal contamination of foods.

2.6.5 Control

As *E. coli* is a common faecal contaminant, there are no specific control measures peculiar to this organism; instead we need to consider general guidelines. Special emphasis needs to be placed on avoiding direct faecal contamination of food by stressing strict personal hygienic practices, especially by food-handlers. It is also desirable to reduce the possible hazards of indirect faecal contamination, firstly by avoiding sewage-contaminated water, and secondly by thorough cooking of all potentially contaminated foodstuffs. Obviously, the avoidance of cross-contamination between raw and cooked foods is of the utmost importance. Finally, it should always be emphasized that untreated sewage should under no circumstances be used in the fertilization of crops or vegetables.

3 Toxic bacterial food poisoning

A.R. Eley

Bacterial food poisoning may be due to actual infection of the intestine (Chapter 2); it may also be caused by the action of toxins produced by the organisms either in the food before consumption or in the intestine (Table 3.1).

This chapter will deal with those organisms responsible for pre-formed toxin in foods, for example, *Staphylococcus aureus*, *Clostridium botulinum* and *Bacillus cereus* (emetic), and those which form toxin in the intestine, for example, *Clostridium perfringens*, *B. cereus* (diarrhoeal), enterotoxigenic *Escherichia coli* (ETEC), and enterohaemorrhagic *Escherichia coli* (EHEC) (Table 3.2).

Other bacteria such as shigella, plesiomonas and aeromonas which are occasionally associated with food poisoning and may produce toxins will be discussed in Chapter 4.

3.1 STAPHYLOCOCCUS AUREUS

These are Gram-positive, facultatively anaerobic, non-spore-forming cocci which are coagulase and deoxyribonuclease (DNase) positive.

3.1.1 Pathogenesis

Staphylococcal food poisoning is caused by the ingestion of food containing pre-formed toxins secreted by the bacteria. These are known as staphylococcal enterotoxins, and eight serologically distinct types (A, B, C_1, C_2, C_3, D, E and F) have so far been recognized. Enterotoxin F has now been shown to be identical biochemically to toxic shock syndrome toxin 1 (TSST-1) which produces toxic shock syndrome commonly associated with the use of tampons during menstruation.

Experiments in primates on the pathogenesis of staphylococcal food poisoning have shown that these toxins cannot be considered as classical enterotoxins like cholera toxin, since they do not act directly on intestinal cells. Instead, the

Table 3.1 Properties of toxins produced by food poisoning bacteria

	Molecular weight (Da)	Genetic location	Proteinaceous	Heat stability*	Neurotoxicity	Increases cGMP/cAMP	Action unknown
Staph. aureus							
A, B, C₁, C₂, C₃, D, E, F	28–35 000	Chromosomal	✓	✓	✓	✗	✗
C. botulinum							
A, B, E, F, G	150 000	Phage	✓	✗	✓	✗	✗
B. cereus (E)	5000	?	✗	✓	✗	✗	✓
Cl. perfringens (A)	35 000	?	✓	✗	✗	✗	✓
B. cereus (D)	50 000	?	✓	✗	✗	✓	✗
ETEC							
STₐ	2000	Plasmid	✓	✓	✗	✓	✗
STᵦ	5000	Plasmid	✓	✓	✗	✗	✓
LT	86 000	Plasmid	✓	✗	✗	✓	✗
EHEC							
VT1	29–31 000 / 5–6 000	Phage	✓	?✗	✗	✗	✓
VT2	32 500 / 5–6 000	Phage	✓	?✗	✗	✗	✓

*Heat-stability: 100°C for 30 minutes

Table 3.2 Basic characteristics of toxic bacteria causing food poisoning

	Gram-reaction	Cell morphology	Catalase	Oxidase	Motility at 37°C	Growth at 4°C
Staphylococcus aureus	+	coccus	+	−	−	−
Clostridium botulinum	+	bacillus	−	−	+	v
Bacillus cereus						
(emetic and diarrhoeal)	+	bacillus	+	+	+	v
Clostridum perfringens	+	bacillus	−	−	−	−
Escherichia coli (ETEC,						
EHEC)	−	bacillus	+	−	v	−

v, variable reaction.

toxins act on receptors in the intestine, with the stimulus reaching the vomiting centre in the brain via the vagus nerve, and should, therefore, be considered as neurotoxins.

The toxins are produced during active growth of the bacteria in foods, often during storage. Each toxin is a single polypeptide chain which is resistant to many proteolytic enzymes and generally withstands boiling for up to 30 minutes (i.e. is heat-stable), although the vegetative cells would not survive such conditions. Hence, if toxin is allowed to be produced in food, although later cooking may kill the bacteria, toxin activity may remain intact.

The type of enterotoxin most frequently involved in food poisoning is staphylococcal enterotoxin A (SEA) which is found in the food associated with approximately 75% of outbreaks due to this organism. SED is the second most important cause of food poisoning outbreaks. Previous studies suggested an association between the enterotoxin type, certain foods and the source of the staphylococci (e.g. human skin). However, more recent evidence has revealed that organisms producing enterotoxins other than SEA may be found more frequently in clinical specimens than was originally thought. Generally, approximately 15–20% of Staph. aureus isolates from humans are enterotoxigenic; this explains the importance of the food-handler in transmission (section 3.1.4).

3.1.2 Clinical features and prognosis

This type of food poisoning is characterized by nausea, vomiting, abdominal pain and prostration, often with diarrhoea but without fever, approximately 1–6 hours after ingestion of contaminated food (Table 3.3).

Although most patients usually recover completely within 24 hours

Table 3.3 Clinical features of the illnesses produced by the major toxin-forming food poisoning bacteria

	Staph. aureus	Cl. botulinum	B. cereus (E)	Cl. perfringens	B. cereus (D)	ETEC	EHEC
Incubation time (h)	1–6	12–35	1–6	8–24	8–16	12–72	48–288
Duration of illness (h)	6–24	(1 week–6 months)	6–24	12–24	12–24	24–168	72–192
Vomiting	+	±	+	–	–	–	–
Nausea	+	±	+	±	+	+	+
Diarrhoea	+	±	±	+	+	+	+
Abdominal pain	+	±	+	+	+	+	+
Fever	–	–	–	–	–	+	–

–, rarely seen; ±, sometimes seen; +, often present; (E), emetic; (D), diarrhoeal.

without specific therapy and show a low overall case–fatality ratio, the presenting symptoms are serious enough for up to 10% of cases to be admitted to hospital.

3.1.3 Incidence and epidemiology

This is the second most common cause of food poisoning in the USA and has been reported to be prevalent in Hungary. In both countries the frequency of outbreaks is thought to be linked to dietary habits. It is suggested that in the USA the incidence of disease may be connected with the widespread consumption of commercially prepared foods and the frequency of large communal meals which would result in outbreaks. From the figures recorded, *Staph. aureus* food poisoning appears to be relatively uncommon in the UK and Japan. However, this is probably due to the rapid recovery rate which accounts for the disease being infrequently reported.

Staph. aureus is almost always transmitted to food from a human source, e.g. food-handlers, or by cross-contamination from another source (such as utensils) previously contaminated by humans. However, an occasional source of infection has been shown to be dairy produce especially from cattle, sheep and goats. Human-to-food transmission is of particular note in epidemiology, as between 25% and 50% of the population may be carriers of *Staph. aureus*. Fortunately, most of the strains responsible for food poisoning (enterotoxin-producers) belong to types within phage group III, and so the spread of infection can be more easily monitored than if they all belonged to different phage groups. Phage typing (section 2.1) of staphylococci has been used for many years as an epidemiological tool in tracing the source of contamination in food poisoning outbreaks. Checking food poisoning-associated staphylococcal isolates for enterotoxin production is also useful. However, it is not always possible to trace the source of the organism by enterotoxin production alone, and in this case the combination of phage typing and characterizing enterotoxin production is very important.

Traditionally, the genus *Staphylococcus* is subdivided into two groups of strains on the basis of the coagulase test (based on the ability to coagulate plasma). The vast majority of food poisoning strains are *Staph. aureus* (coagulase-positive). However, at least one outbreak has been reported due to *Staph. epidermidis* (coagulase-negative) which normally does not produce an enterotoxin. Therefore, coagulase-negative staphylococci should not be ignored, especially if they are present in large numbers in food. However, it should be remembered that *Staph. epidermidis* is a normal skin commensal and its presence in food may reflect poor levels of hygienic practices. Moreover, enterotoxin production in this organism has been poorly studied.

More recently, two new species, *Staph. intermedius* and *Staph. hyicus*, which may be positive in tests for coagulase and DNase, have been found to produce

Table 3.4 Characteristics of *Staphylococcus* species

Property	Staph. aureus	Staph. epidermidis	Staph. intermedius	Staph. hyicus
Coagulase	+	−	+	v
DNase	+	−	+	v
Pigment	+	−	−	−
Haemolysis on blood agar	+	v	+	−
Mannitol fermentation	+	−	−	−
Hyaluronidase	+	−	−	+

v, variable reaction.

enterotoxins (Table 3.4). However, at present their significance as causes of food poisoning is unknown.

3.1.4 Ecology and foodstuffs

Staph. aureus is a major pathogen of humans, causing a wide range of infections, and can be found as a commensal on the skin and in the anterior nares of the nose. This organism may also be present in the air, in milk and in sewage.

Foods commonly implicated in this type of food poisoning include cooked foods eaten cold, e.g. processed meats and eggs, and prepared foods such as custards and other dairy produce (Table 3.5). The former are important as enterotoxin production is more likely when competing organisms are absent or few, as in a cooked food.

3.1.5 Control

One of the major problems associated with the control of *Staph. aureus* food poisoning is the high carriage rate of the organism in humans, which makes for a significant risk of contamination by food-handlers. Ideally, food should be cooked and eaten immediately after handling, to prevent the growth of contaminating bacteria. Keeping food at ambient temperature or above before heating allows the formation of toxin, which may not then be destroyed by normal cooking. Clearly, it is not always possible to avoid a delay between handling and cooking; in addition, prepared and processed foods which are eaten cold have to be handled after the cooking process is complete. Care should therefore be taken to minimize handling and to keep foods refrigerated before cooking or serving.

Table 3.5 Foodstuffs commonly associated with bacterial causes of toxic food poisoning

Bacteria	Meat and/or meat products	Poultry	Eggs	Milk	Dairy products	Seafood and/or shellfish	Vegetables	Cereals
Staph. aureus	✓	–	✓	–	✓	–	–	–
Cl. botulinum	✓	–	–	–	–	✓	✓	–
B. cereus (E)	–	✓	–	–	–	–	–	✓
Cl. perfringens	✓	–	–	–	✓	–	–	–
B. cereus (D)	✓	–	–	–	–	–	–	–
ETEC	✓	–	–	–	–	–	–	–
EHEC	✓	–	–	✓*	–	–	–	–

*Unpasteurized milk.

3.2 *CLOSTRIDIUM BOTULINUM*

Clostridia are Gram-positive, obligately anaerobic, endospore-forming bacilli.

3.2.1 Pathogenesis

Botulism is caused by the production of botulinum toxins, which are protein-aceous neurotoxins and are the most potent natural poisons known. The amount of toxin which may be described as a fatal dose has been estimated from clinical cases and animal studies to be between 0.1 and 1.0 μg. Seven distinct types of botulinum toxin are recognized, although botulism in humans is usually caused by types A, B and E, and more rarely F and G. Once the toxin is absorbed, it attaches to the neuromuscular junction of affected nerves and prevents the release of acetylcholine, thus causing muscular paralysis. In severe cases the paralysis can be profound, with death resulting from respiratory failure within 24 hours.

In food-borne botulism, the food item becomes contaminated with spores from the environment, which are not destroyed by the initial cooking or processing. If the food is then kept in conditions appropriate for growth, the spores may germinate, leading to production of toxin. If not destroyed by heating before serving, the toxin can be ingested with the food item and absorbed. Fortunately, the purified toxin is heat-labile, and is inactivated by heating at 80°C for 10 minutes. Of note is the ability of this organism not only to grow at low temperature and/or low pH but also to produce toxin under these adverse conditions. In certain strains growth has taken place at 4°C and at a pH value of 4.0.

3.2.2 Clinical features

The first symptoms of botulism usually develop between 12 and 36 hours after consumption of contaminated food. However, in certain instances the incubation period may be as long as 8 days depending, in part, on the dose of toxin ingested. Furthermore, symptoms may differ and are related to the causative *Cl. botulinum* type. Symptoms include dizziness, difficulty in swallowing, slurred speech, weakness of limbs and blurred or double vision. Although nausea, vomiting and diarrhoea may be reported there is usually no fever. Breathing problems are caused by respiratory paralysis which may lead to death by asphyxiation.

The differential diagnosis of botulism includes cerebrovascular disease, post-infective neuropathies of Guillain–Barré or Miller–Fisher type, and myasthenia gravis.

3.2.3 Prognosis

Botulism is a serious disease but fatalities may be considerably reduced if polyvalent botulinum anti-toxin is administered at an early stage, and patients are provided with some form of respiratory support such as positive-pressure ventilation in an intensive care unit. Before 1949, the case–fatality rate of botulism was approximately 60%, but it has since declined and is currently less than 10%. With the correct treatment only one patient died out of a total of 27 cases in the 1989 UK hazelnut purée outbreak.

3.2.4 Incidence and epidemiology

Food-borne botulism is a disease of humans and animals. In humans it is seen infrequently in the UK, but is more commonly found in the USA, especially Alaska, and in Asia (Table 3.6).

Between 1922 and 1989 fewer than 30 cases of botulism were recorded in the UK; the largest outbreak ever recorded in this country occurred in 1989, and involved 27 patients (Table 3.7). The contaminated food was canned hazelnut purée which was added to yoghurt and consumed in north-west England and North Wales. In China and Alaska, where dietary practices differ, over 1000 outbreaks have occurred over the past 30 years. In these areas, botulism is associated with traditional foods such as home-processed fish or preserved bean curd which undergo fermentation and are consumed uncooked.

There is considerable variation in the geographical distribution of toxins responsible for diseases, and this usually correlates with identification in soil of organisms that produce specific toxin types. In China, for example, type A was the toxin most frequently isolated from patients in the north-west region; in the north type B predominates, while type E is most common in the north-eastern region. In a 30-year study in Alaska type E accounted for 73% of laboratory confirmed outbreaks. In the recent UK hazelnut purée outbreak the toxin responsible was type B.

Table 3.6 Recorded food-borne botulism – worldwide

Country	Period	Outbreaks	Total cases	Fatal cases
USA	1971–1992	338	795	75
Canada	1971–1989	79	202	28
Germany	1983–1989	96	206	10
France	1978–1992	235	406	10
UK	1978–1992	3	32	3
Belgium	1982–1989	11	25	1
Poland	1984–1987	1301	1971	46

In most instances the implicated food is prepared at home. (Source: Hauschild, A.H.W. (1992), *Clostridium botulinum – ecology and control in foods*, (eds. Hauschild A.H.W. and Dodds K.L.), p. 70, published by Marcel Dekker, USA, 1992.)

Table 3.7   Food-borne botulism in the UK

Year	No. of cases	No. of deaths	Food vehicle	C. botulinum type
1922	8	8	Duck paste	A
1932	2	1	Rabbit and pigeon broth	
1934	1	0	Jugged hare	
1935	5?	4?	Vegetarian nut brawn	A
1935	1	1	Minced meat pie	B
1947	5	1	Macaroni cheese	
1955	2	0	Pickled fish from Mauritius	A
1978	4	2	Canned salmon from US	E
1987	1	0	Kosher airline meal	A
1989	27	1	Hazelnut yoghurt	B

(Source: Gilbert, R. (1990), *Clinical and Molecular Aspects of Anaerobes*, (ed. Borriello, S.P.), p.88.)

Two other types of botulism, known as wound botulism and infant botulism, have been identified. Wound botulism is caused by *Cl. botulinum* spores contaminating wounds where they germinate, leading to neurointoxication. Infant botulism is a newly recognized form of botulism in which ingested spores of *Cl. botulinum* germinate and multiply in the intestine of the patient, producing toxin *in vivo*. Recently, there have been reports of several cases in which botulism in an adult was due to the synthesis of toxin within the intestine during an intestinal infection with *Cl. botulinum* (cf. infant botulism).

3.2.5 Ecology and foodstuffs

Cl. botulinum may be found in the intestines of humans and other animals, and in soil and mud, from where it can contaminate vegetables.

As *Cl. botulinum* is strictly anaerobic, botulism is only associated with foods that can provide suitable anaerobic conditions. *Cl. botulinum* spores are heat-resistant (they can survive for 2 hours in boiling water), and are only killed under proper food processing conditions. If such conditions are not maintained, the spores may germinate and allow vegetative cells to emerge, thereby releasing the toxin.

Traditionally, botulism has been associated with home-preserved foods and vegetables, although this source has decreased in importance over the past few years. Nowadays, the disease is more closely associated with improperly processed canned meats, and traditional fermented foods, such as those made with contaminated vegetables. Contaminated foods are all the more dangerous because they often show no obvious signs of deterioration.

3.2.6 Control

The usual methods for controlling *Cl. botulinum* in preserved foods are designed to inhibit rather than destroy the organisms, since a major property of spores is their inherent resistance. Food treatments may involve one or more of the following factors: pH, a_w (water activity), refrigeration, salt, Eh (redox potential), nitrite, canning and smoke. Although a number of preserved foods are potentially hazardous, growth of *Cl. botulinum* and toxin production usually occur as a result of faulty manufacturing conditions or through storage at the incorrect temperature. Fortunately, botulinum toxin is destroyed in a few minutes by boiling; so adequate cooking before consumption will render contaminated food harmless.

Since food-borne botulism can, in theory, lead to very large outbreaks, early recognition of cases is essential. Food vehicles can then be identified through epidemiological investigations, products can be withdrawn from sale and appropriate publicity can be given in the media.

3.3 BACILLUS CEREUS (EMETIC)

These are Gram-positive, facultatively anaerobic, endospore-forming bacilli.

3.3.1 Pathogenesis

Although *B. cereus* has been known as an agent of food poisoning since the beginning of the century, it is only recently that we have recognized the existence of two different syndromes, each caused by a different toxin: the emetic form discussed here and the diarrhoeal type described in section 3.5.

The emetic disease follows the ingestion of food contaminated with a low-molecular weight toxin. Little is known about this toxin; it is probably a heat-stable protein associated with and/or produced during spore formation. Its mode of action is thought to be similar to that of the enterotoxin produced by *Staph. aureus* (section 3.1.1).

In outbreaks associated with Chinese 'take-away' dishes, the pattern of toxin production is as follows: spores in rice, which can be quite heat-resistant, are not always killed during the cooking process, which in turn selects for spores of greater heat resistance. When the rice cools and following spore germination, vegetative cell growth may be rapid, especially at room temperature. Of the now large number of vegetative cells in the food, some may sporulate and lead to toxin formation, especially if the rice is left for more than a few hours in a fairly warm atmosphere. Although the rice may be fried before serving, the toxin is heat stable and can withstand exposure at 121°C for 90 minutes.

3.3.2 Clinical features and prognosis

Symptoms often resemble staphylococcal food poisoning, with rapid onset of nausea, vomiting and malaise, usually within one to six hours after consumption of the contaminated food (Table 3.3).

As with staphylococcal food poisoning, complications are rare and recovery is usually complete within 24 hours.

3.3.3 Incidence and epidemiology

In the UK and the USA, the incidence of B. cereus food poisoning is not particularly high in comparison with salmonella and Cl. perfringens (Table 1.2), but the emetic type of the disease is almost always associated with rice and fried rice. Many of the outbreaks have occurred in restaurants serving oriental foods. To underline the importance of rice in the causation of disease, it should be noted that between 1971 and 1978 there were 110 reported incidents of emetic B. cereus food poisoning in the UK, of which all but two were associated with rice, usually in the form of Chinese fried-rice dishes.

Emetic-toxin-producing strains of B. cereus have been characterized by serotyping and are commonly serotypes H.1, H.3 and H.8.

Recently, another Bacillus species, B. subtilis, has also been implicated as a cause of the emetic syndrome, with a very rapid onset. It is presumed that the toxin produced by this organism is similar to that elaborated by B. cereus. B. subtilis food poisoning has been associated mainly with meat and pastry dishes such as sausage rolls, meat pasties and pies.

3.3.4 Ecology and foodstuffs

Apart from being an important cause of food poisoning, B. cereus can also be an opportunist pathogen with the potential for causing severe disseminated infections. In addition, it is a common contaminant, found in large numbers in natural, domestic and hospital environments.

Cereals, especially rice, are implicated in the majority of cases of B. cereus food poisoning. Other food vehicles have included pasta, milk pudding and pasteurized cream.

3.3.5 Control

Since Bacillus species are ubiquitous in the environment, these organisms will almost always contaminate our food. It is only when large numbers of organisms are present or when toxin is produced that B. cereus becomes a hazard. To avoid this we must control spore germination and prevent vegetative cells from multiplying in cooked foods. Cell multiplication during inadequate cooling of cooked cereal-based goods, e.g. rice, is the greatest problem. Ideally, control measures should include a ban on the storage of rice after cooking, or, if this is impractical, rapid cooling and final refrigeration of the product.

3.4 *CLOSTRIDIUM PERFRINGENS*

These are Gram-positive, obligately anaerobic, endospore-forming bacilli, although most strains do not form spores in culture. In contrast to the strains which cause gas gangrene, many strains associated with food poisoning are non-haemolytic on horse-blood agar. As with spores of *Bacillus* species, spores of *Cl. perfringens* will often achieve optimal germination only if they are mildly heated.

3.4.1 Pathogenesis

When contaminated food (typically meat) is cooked, the heat drives off the dissolved oxygen and induces sporulation of the bacteria. As cooling occurs, the spores germinate and vegetative cells continue to multiply, unless the meat is cooled rapidly and kept refrigerated until being reheated thoroughly before consumption. This organism has the ability to proliferate over a wide temperature range (15–50°C) and grows optimally at between 43°C and 47°C, when the mean generation time (the time taken for the number of bacterial cells to double) can be as low as 12 minutes. If a product containing a large number of vegetative cells (> 10^6) is consumed, the organisms multiply when they reach the gastrointestinal tract; this is followed by sporulation in the small intestine, with the subsequent release of enterotoxin.

This enterotoxin was thought to be a sporulation-specific gene product liberated from the sporangium (the outer coat of the spore) on lysis. More recently, enterotoxin synthesis has been detected in non-sporulating cultures of *Cl. perfringens* and it has been claimed that toxin production is not associated with the spore coat. It is now suggested that sporulation and enterotoxin production are coincidental events which occur under similar environmental conditions.

Although its mode of action is not fully established, the enterotoxin has been purified, and it is known to be a protein with a molecular weight of 35 000 Da. The toxin damages epithelial cells on the villi tips and inhibits the absorption of glucose, which causes an efflux of Na^+, Cl^- and water. The result of this is excess fluid movement into the lumen of the gut, leading to diarrhoea.

Numerous exotoxins are produced by *Cl. perfringens*, and the relative production of the four major lethal toxins α, β, γ and σ is used as a basis for dividing the species into five toxin types, A to E. Only the type A strains (which can be further subdivided into a number of serotypes) are responsible for food poisoning.

Enteritis necroticans, also known as necrotizing enteritis or 'pig bel', is a rare but often fatal haemorrhagic disease caused by ingestion of food (often pork) contaminated with *Cl. perfringens* type C. The organisms adhere to the cells of the intestinal wall and produce a necrotizing b-toxin which causes necrosis of the mucosa. If untreated, the disease progresses to gangrene of parts of the small intestine, with shock and severe toxaemia. Recently, necrotizing enteritis associated with *Cl. perfringens* type A has also been reported.

3.4.2 Clinical features and prognosis

Symptoms are abdominal pain and diarrhoea, usually between 8 and 24 hours after the ingestion of contaminated food. Although there may be prostration in debilitated patients, it is not accompanied by fever or vomiting.

The duration of the illness is relatively short, with symptoms disappearing within 12 to 24 hours. Since the disease is usually self-limiting, only supportive therapy is needed, although fatalities do occur occasionally among elderly or debilitated patents, usually through dehydration.

3.4.3 Incidence and epidemiology

This organism is an important pathogen of humans which is responsible for gas gangrene and cellulitis as well as food poisoning.

Cl. perfringens has been implicated in the causation of gastrointestinal disturbances since the 1890s, and is recognized as a common cause of food poisoning throughout the world. In the 1990s *Cl. perfringens* is still the second most frequent cause of reported food poisoning in the UK and the third most frequent in the USA. Worldwide, serotype 1 is most commonly reported.

Outbreaks are often associated with large-scale catering of the type found in canteens, schools, hospitals and other large institutions. Problems can arise with advance preparation of large quantities of food such as stew, which are difficult to cool rapidly before storage. Slow cooling will result in the rapid multiplication of vegetative cells.

Enteritis necroticans ('pig bel') was first reported in 1949 in north-west Germany, where it was called Darmbrand. It is endemic in Papua New Guinea; sporadic cases have occurred worldwide. The incidence of 'pig bel' in Papua New Guinea has been reduced by vaccination using a toxoid preparation of type C cultures of *Cl. perfringens*.

3.4.4 Ecology and foodstuffs

Cl. perfringens is commonly found in the intestines of humans and animals and in the soil, resulting in the contamination of meats and vegetables respectively.

Cooked meat, poultry, fish, stews, pies and gravies all provide excellent conditions for growth.

3.4.5 Control

In the majority of outbreaks, the prime factor is the failure to refrigerate cooked foods properly, especially when they have been prepared in large quantities. If foods are not heated rapidly and uniformly, spores of *Cl. perfringens* may survive the cooking process and then germinate when a suitable temperature is reached during cooling. Thorough reheating of refrigerated foods is essential to kill any vegetative cells present. Preventive measures, therefore, depend on food-handlers having adequate knowledge of proper food preparation and storage techniques.

3.5 *BACILLUS CEREUS* (DIARRHOEAL)

3.5.1 Pathogenesis

In addition to the emetic syndrome described in section 3.3, *B. cereus* is also responsible for a quite distinct diarrhoeal syndrome, caused by a diarrhoeal toxin which is formed as follows. Since spores are more heat-resistant than vegetative cells, they can survive boiling in foods such as stews, and these spores can germinate in the gastrointestinal tract after consumption of the contaminated food. Toxin is produced on germination.

This toxin is a protein enterotoxin with a molecular weight of approximately 50 000 Da, which is thought to act in a similar way to cholera toxin, causing fluid secretion in the gut by activating adenylate cyclase (Table 3.1). The toxin from *B. cereus*, however, is heat-labile and its activity is considerably reduced above 45°C.

3.5.2 Clinical features and prognosis

This disease is characterized by an incubation period of 8–16 hours followed by abdominal pain, cramps and a profuse watery diarrhoea; vomiting and fever are rarely seen (Table 3.3).

Recovery is usually complete within 24 hours. If there is severe diarrhoea, certain at-risk groups such as the debilitated and the elderly should be observed carefully for dehydration.

3.5.3 Incidence and epidemiology

This diarrhoeal syndrome is similar to that caused by *Cl. perfringens* and has been reported widely from Northern and Eastern Europe as well as the UK, USA, Canada and Japan. In the UK this illness is much less common than the emetic type, whereas in Hungary between 1960 and 1968 the diarrhoeal syndrome was the third most common cause of food poisoning. Current trends worldwide indicate that the number of outbreaks of the diarrhoeal syndrome has declined and is declining still further.

Diarrhoeal-toxin-producing strains can be characterized by serotyping but because few outbreaks have been studied in this way, it is too early to be able to identify the most common serotypes with certainty.

More recently, B. *licheniformis* has been reported as a cause of a similar diarrhoeal syndrome, although a higher incidence of vomiting was noted. The associated foodstuffs were as for *B. cereus*, but also included minced beef, curries and pâtés.

3.5.4 Ecology and foodstuffs

The ecology of *B. cereus* (diarrhoeal) is the same as for *B. cereus* (emetic) (section 3.3).

A wide variety of foods, including meat and vegetable dishes, soups, stews, sausages, sauces and desserts have been implicated in the diarrhoeal syndrome.

3.5.5 Control

See section 3.3 for control of *B. cereus* (emetic) and section 3.4 for control of *Cl. perfringens*.

3.6 *ESCHERICHIA COLI*

Although certain types of *E. coli* such as EPEC and EIEC have already been described in Chapter 2, other groups which produce specific, distinct toxins have been implicated as causes of food poisoning. These two groups are known as enterotoxigenic (ETEC) and enterohaemorrhagic (EHEC) and are associated with certain serotypes (Table 3.8).

3.6.1 Enterotoxigenic *E. coli*

(a) Pathogenesis

After ingestion, bacteria that survive the hostile environment of the stomach must penetrate the mucous layer of the small intestine, where they adhere to mucosal cells. There they produce either or both of two types of enterotoxin, LT (heat-labile) and ST (heat-stable), which may result in a profuse watery diarrhoea; typically this is less severe than in patients with cholera.

Table 3.8 Major O serotypes of ETEC and EHEC
associated with diarrhoeal disease

Enterotoxigenic
    06, 08, 020, 025, 027, 063, 078, 080, 085, 0115
    0128ac, 0139, 0148, 0153, 0159, 0167
Enterohaemorrhagic
    01, 026, 091, 0111, 0113, 0121, 0128, 0145, 0157

LT has a molecular weight of 86 000 Da, resembles cholera toxin structurally and antigenically, and has a similar mode of action (i.e. it increases adenylate cyclase activity). Two types of ST have been described, ST_a and ST_b, with a molecular weight of 2000 and 5000 Da, respectively. ST_b is at present poorly characterized and its mode of action is uncertain, although it is thought to be complex. The mode of action of ST_a is not fully understood, but its binding to a receptor leads to increased guanylate cyclase activity with conversion of GTP to cyclic GMP. The latter acts as a messenger like cAMP to cause a physiological response (section 1.3).

(b) Clinical features and prognosis

After an incubation period of between 12 and 72 hours symptoms include diarrhoea, which may be quite severe, fever, and abdominal pain. Although nausea is common, vomiting is rarely seen.

This illness may well last for 1–7 days and is usually self-limiting. However, in poorly-nourished children in developing countries, gastroenteritis may last for weeks and may cause severe malnutrition. If diarrhoea is severe or persists for a long period of time rehydration may have to be considered.

(c) Incidence and epidemiology

ETEC are responsible for producing travellers' diarrhoea, typically in individuals travelling from areas of good hygiene and temperate climate to ones with lower hygienic standards, such as developing countries. The disease may be food- or water-borne. The incidence of ETEC food poisoning in the UK and USA is relatively low. Its incidence elsewhere is less well known as the organism is quite difficult to recognize in mixed culture (*E. coli* is a common gut commensal), and skilled personnel and technology are required to identify it fully. Briefly, isolation techniques involve screening *E. coli* colonies for the serotypes associated with enterotoxin production (Table 3.8).

(d) Ecology and foodstuffs

These organisms can be isolated from many raw foods, especially those of animal origin. As *E. coli* is part of the intestinal flora in humans, its presence may also be a sign of unhygienic food handling and can serve as an indicator of faecal pollution.

(e) Control

See section 2.6.5.

3.6.2 Enterohaemorrhagic *E. coli*

(a) Pathogenesis

Disease is due to the production of one or more phage-encoded verocytotoxins (VTs) called VT1, VT2 and VT3. These toxins are closely related to shiga toxin produced by *Shigella dysenteriae* serotype 1, the causative organism of bacillary dysentery. Verocytotoxin is so-called because it was detected by its cytopathic effect on African green monkey kidney (i.e. Vero) cells cultured *in vitro*.

Pathological effects include morphological changes in epithelial cells, increased mitotic activity in crypts, mucin depletion and an infiltration of polymorphonuclear cells into the mucosa. These changes are always associated with the presence of free VT in the colon and result in either watery and/or bloody diarrhoea.

The infective dose is thought to be low and is estimated to be fewer than 100 cells.

(b) Clinical features

Typically haemorrhagic colitis (HC) is produced. This is a diarrhoeal disease characterized by copious, overt blood in the stool accompanied by severe abdominal pain. The latter may be so intense as to resemble appendicitis and lead to some patients undergoing a laparotomy. Vomiting may occur but there is little or no fever. Approximately 10% of cases usually develop into a serious illness called haemolytic uraemic syndrome (HUS), which consists of a triad of symptoms: acute renal failure, thrombocytopaenia and microangiopathic haemolytic anaemia. These symptoms may require the patient to undertake dialysis and can lead to permanent kidney damage. An unusual feature of EHEC is the wide variation in incubation period, which usually ranges from 2–8 days, but may be as long as 12 days.

(c) Prognosis

Although most patients have a self-limited illness of approximately 8 days' duration, deaths have been reported in elderly patients with underlying medical problems. In children, especially, one should be aware of the potential progression of the symptoms of HC into HUS. Treatment is a problem, as there is little evidence to show that antimicrobial therapy is effective. However, in serious infections ciprofloxacin is the drug of choice.

(d) Incidence and epidemiology

EHEC was first reported as a cause of HC in the USA in 1982. Subsequently, outbreaks and sporadic cases have been reported in North America (especially Canada), in Japan and in the UK (see Table 8.4). It is now thought that between 10 000 to 20 000 E. coli 0157:H7 infections occur each year in the USA. Sixteen outbreaks were reported in 1993 and another 11 during the first 6 months of 1994. As E. coli 0157 has been isolated only rarely from food its epidemiology therefore remains unclear.

Initially, outbreaks of HC were associated with one serotype, 0157:H7; it is now known that other serotypes, e.g. 026, 0111, 0113, 0121 and 0145 may also cause HC or less severe diarrhoea.

(e) Ecology and foodstuffs

In the UK the first report of isolation of E. coli 0157 from bovine faeces was in 1989 and this indicated that cattle may be the possible reservoir of infection. Subsequently it was shown that contamination of carcasses during slaughter and processing may be how beef and beef products become contaminated. Recently, the first incident of E. coli 0157 infection in the UK was described in which a suspect food source (milk) was confirmed microbiologically, and this represented a further means by which the organism may be transmitted from cattle to humans. Similar findings implicating raw or undercooked beef or raw milk have also been reported from North America.

(f) Control

Consumers should avoid eating raw or partially cooked foods of animal origin especially ground beef, and take particular care with procedures which may lead to cross-contamination from raw to cooked foods.

Because of the severity of the diseases caused and a significant increase in incidence in the USA, in 1994 The American Gastroenterological Association Foundation published a number of recommendations to help combat E. coli 0157. Some of these recommendations included more active surveillance programmes in outbreaks and routine testing of all stool specimens for this organism. The latter would entail culturing onto sorbitol MacConkey medium and testing suspect colonies for agglutination with a serum containing antibody to the flagellar antigen H7. This is useful as, unlike most other E. coli, serotype 0157 does not ferment sorbitol and can be easily differentiated.

4 Other bacterial pathogens

A.R. Eley

Bacterial food poisoning has been discussed in Chapters 2 (Infective bacterial food poisoning) and 3 (Toxic bacterial food poisoning), where organisms were divided into two groups according to whether infectivity or toxin production was considered to be the major mode of pathogenesis. The majority of bacterial pathogens known to cause food poisoning are found within these two groups. There is, however, a third group of bacteria implicated in food poisoning, which may be termed minor pathogens, either because the total number of cases recorded per year is small or because their pathogenicity in this context is generally regarded as low; it is with these organisms that this chapter will be concerned. A summary of their basic characteristics is given in Table 4.1.

4.1 LISTERIA MONOCYTOGENES

These are Gram-positive, facultatively anaerobic, non-spore-forming bacilli or cocco-bacilli sometimes occurring in 'Chinese letter' arrangements on Gram stain.

4.1.1 Pathogenesis

Although the human intestine is the portal of entry for *L. monocytogenes* associated with food, the common clinical picture of listeriosis is of meningitis and/or

Table 4.1 Basic characteristics of other food poisoning bacteria

	Gram reaction	Cell morphology	Catalase	Oxidase	Motility at 37°C	Growth at 4°C
Listeria monocytogenes	+	bacillus	+	−	−*	+
Aeromonas spp.	−	bacillus	+	+	+	v
Shigella spp.	−	bacillus	+	−	−	−
Plesiomonas shigelloides	−	bacillus	+	+	+	−
Streptococcus pyogenes	+	coccus	−	−	−	v
Enterococcus faecalis	+	coccus	−	−	v	v

*Motile at 20–25°C; v, variable reaction.

Table 4.2 Differentiation of *Listeria* species

Biochemical test	Listeria					
	monocytogenes	ivanovii*	innocua	welshimeri	seeligeri	grayi
Glucose	+	+	+	+	+	+
Aesculin	+	+	+	+	+	+
Rhamnose	+	−	v	v	−	−
Xylose	−	+	−	+	+	−
Mannitol	−	−	−	−	−	+
Hippurate hydrolysis	+	+	+	+	+	−
Beta-haemolysis	+	+	−	−	+	−
Nitrate reduction	−	−	−	−	−	v
H2S by lead acetate strip	−	−	−	−	−	+

v, variable reaction.
*The two subspecies of *L. ivanovii* are differentiated on the basis of acid production from ribose (subsp. *ivanovii*) and N-acetyl-β-D-mannosamine (subsp. *londoniensis*)

septicaemia. Host resistance to listeria infection is thought to be important in determining whether disease occurs, hence the association of listeriosis with the unborn, newborn, pregnant women and the elderly: in essence, persons whose cell-mediated immunity is thought to be impaired. However, healthy people with no underlying conditions may also become ill with listeriosis, although this is rare. At present the infective dose is unknown, although it is likely to be much reduced when the organism behaves as an opportunist pathogen. For susceptible individuals it has been suggested that it is as low as 100 bacteria.

The most important pathogenic property of virulent *L. monocytogenes* is related to the production of a β-haemolysin called listeriolysin. It has been shown that the loss of this haemolysin is accompanied by a loss of virulence. Other gene products involved in pathogenesis are a protein, internalin, that is involved in the invasion of host cells, phospholipase, a metalloprotease and an actin assembly protein. All the genes encoding for these products appear to be under the control of a gene *prfA*.

A correlation between the production of haemolysin and phospholipase has been observed in *L. monocytogenes*; only virulent strains are lipolytic.

In very rare instances, two other *Listeria* species, *L. seeligeri* and *L. ivanovii* are pathogenic to humans. Both of these strains are also haemolytic, which differentiates them from other saprophytic species of listeria (Table 4.2).

4.1.2 Clinical features

Most healthy persons whose gastrointestinal tract is contaminated with listeria are probably symptomless or suffer only mild symptoms that may well go unnoticed. Where listeria behave as opportunist pathogens of the gastro-intestinal tract causing disease, symptoms may include nausea, vomiting and abdominal pain although they are not often a significant feature of infection, usually before the onset of fever. In pregnant women especially these symptoms may also be accompanied by a flu-like illness. Generally, disease in these women is fairly mild, but the illness in the infant can be much more severe leading to neonatal sepsis or stillbirth. However, it must be stressed that in patients infected by listeria, symptoms are much more variable than with certain other bacteria, e.g. *Campylobacter jejuni* (Table 4.3).

Previously, the incubation period was thought to vary from a few days to a few weeks, but now it is known that in exceptional cases it may be as short as 1 day after the consumption of heavily contaminated food. In more serious instances fever may be accompanied by bacteraemia, which can lead to meningitis, with an overall case fatality rate of about 30%.

4.1.3 Prognosis

Those patients who were previously healthy rarely develop the serious type of listeriosis unless they were exposed either to excessive numbers of viable cells

Table 4.3   Clinical features of the illnesses produced by the other bacterial causes of food poisoning

	Listeria monocytogenes †	*Shigella species*	*Aeromonas species*	*Plesiomonas shigelloides*	*Streptococcus pyogenes/ Enterococcus faecalis*
Incubation time (h)	(1 day–3 weeks)	24–168	18–24	24–168	2–48
Duration of illness (h)	48–72	(1–2 weeks)	24–168	24–168	24
Vomiting	±	−	±	±	±
Nausea	±	±*	±	+	+
Diarrhoea	±	+*	+	+	+
Abdominal pain	±	+*	±	+	±
Fever	+	+*	+	±	−

−, rarely seen; ±, sometimes seen; +, often seen; †, symptoms often very variable;
*, symptoms often less severe with *Sh. sonnei*

of *L. monocytogenes* or to strains of unusually high virulence. Recovery is usually complete within a few days without specific therapy. Those patients with an underlying condition may ultimately predispose to meningitis. For severe listeriosis, antibiotic treatment consisting of ampicillin (erythromycin if allergic to penicillins) and gentamicin should be administered as quickly as possible.

4.1.4 Incidence and epidemiology

Relative to other bacterial causes of food poisoning such as salmonella and campylobacter, the incidence of disease caused by *L. monocytogenes* is low worldwide. In England, Wales and Northern Ireland, for example, only 288 cases of all types of listeriosis (including those not thought to be food-related) were reported in 1988, compared with almost 30 000 cases each of salmonellosis and campylobacter infection. In contrast, the actual incidence of listeria in foods is very similar to that of the classical Gram-negative enteric pathogens (i.e. very high).

In North America and some European countries, including the UK there was a definite increase in listeriosis during the late 1980s. It is possible that this increase was partly due to a rise in the number of persons who are more susceptible to the disease, for example, the immunocompromised. Other reasons may include the greater public awareness of reported cases and improved laboratory techniques and vigilance. However, it is unlikely that even the combined effect of these factors accounted for the steep rise in the total number of listeriosis cases in England, Wales and Northern Ireland (Table

Table 4.4 Human listeriosis in England, Wales and Northern Ireland, 1984–1993

	1984	1985	1986	1987	1988	1989	1990	1991	1992	1993
England and Wales	112	136	129	238	277	237	116	127	106	102
N. Ireland	1	0	2	6	11	5	3	3	2	4
Total	113	136	131	244	288	242	119	130	108	106

(Source: PHLS Communicable Disease Surveillance Centre/Food Hygiene Laboratory.)

4.4) which reached a peak in 1988, but has now dramatically declined.

Worldwide, seven documented large-scale, food-related outbreaks of listeriosis have been recognized since 1981 when it was first confirmed that *L. monocytogenes* was a cause of food poisoning: three in North America, two in France, one in the UK and one in Switzerland. Four involved milk products, two involved pork and the other implicated coleslaw (see Table 8.3). Following these seven outbreaks, a World Health Organization (WHO) working group concluded that *L. monocytogenes* is an environmental micro-organism whose primary means of transmission to humans is through foodstuffs contaminated during production and processing. In some sporadic cases food has been implicated, and recent epidemiological studies have linked uncooked and undercooked food with disease. One of the problems facing epidemiologists is that because of the variable and often long incubation period, disease may not be evident until several weeks after the consumption of infected food, making it difficult to trace the source.

Although *L. monocytogenes* can be divided into several groups by serotyping, this is of limited epidemiological use because the majority of human isolations in the UK belong to serotypes 1/2 and 4.

4.1.5 Ecology

This organism is ubiquitous and is widespread in the environment, where it may survive for long periods. This may be due to a number of factors, perhaps most importantly its capacity for relatively rapid growth at refrigeration temperatures (the lower limit for growth is approximately 0°C). The organism can also survive drying and freezing, and is less sensitive to heat treatment than salmonella and campylobacter. There is still controversy in the literature about the ability of *L. monocytogenes* to survive heat treatment, especially in the pasteurization of milk. It should of course be remembered that pasteurization is not a method of sterilization but a technique for killing certain pathogens and spoilage organisms. Most studies have concluded that although the D-values (time for a 10-fold reduction in viable numbers) are longer than for most other commonly occurring food poisoning bacteria, the temperature routinely adopted for pasteurization over the recommended time-scale would be adequate. The exception is with high-temperature short-time (HTST) pasteurization, which has failed to inactivate *L. monocytogenes* when present in large numbers in samples. However, such numbers of organisms are unlikely to occur in practice, since few animals in any dairy herd would be excreters of *L. monocytogenes*, and there would therefore be a significant dilution effect in normal bulk storage of milk. In addition to some tolerance to acidic conditions, *L. monocytogenes* is also able to survive relatively high osmotic pressures, as shown by its ability to grow in the presence of 10% sodium chloride. Its presence has been reported in animal and human faeces (although there is very considerable variation in the isolation rates), and consequently in sewage; it has also been isolated from slurry, soil, surface water, vegetation and silage.

4.1.6 Foodstuffs

Many cases of listeriosis have not been shown to be directly related to a known food source of *L. monocytogenes*, and it is often extremely difficult to differentiate between cases that are food-related and those that are not.

Many recent surveys have been conducted and have shown the organism to be commonly isolated from raw meats and chilled foods. Of particular note are the high number of *L. monocytogenes* that have been found in certain types of soft cheese and pâté. Table 4.5 shows the incidence of *L. monocytogenes* in foods, indicated by the presence or absence of the bacterium in 25 g of sample.

4.1.7 Control

L. monocytogenes is particularly difficult to control, since it is ubiquitous and widespread in the environment, and because it possesses physiological characteristics (e.g. multiplication at refrigeration temperatures) that allow growth under conditions that are usually adverse for most other pathogenic bacteria. Pasteurization will, however, render contaminated milk fit for human consumption.

For the consumer perhaps one of the most significant effects of the contemporary listeria 'scares' has been the emphasis on the need for low refrigeration temperatures and their careful monitoring. For chill-storage the temperature

Table 4.5 The incidence of *L. monocytogenes* in foods*

Food	No. of samples	Incidence (%)
Raw beef	29	38
Raw pork	33	36
Raw pork sausages	25	52
Pâtés	73	51
Raw chicken	100	60
Raw milk	137	4
Pasteurized milk	41	0
Unpasteurized cheese (camembert/brie)	18	55
Pasteurized cheese	51	0
Ice-cream	394	1
Frozen seafood products	57	26
Pre-packed salads	60	7
Fresh-cut vegetables (salads)	25	44
Potatoes	132	21

*After Lund, B.M. (1990) *British Food Journal*, **92**, 13–22.

range of 0–4°C is to be preferred, as even a small bacterial inoculum held at these temperatures may outgrow competing organisms and produce a dose sufficient to infect susceptible individuals and cause disease.

Initially, measures that minimize contamination of the raw product should be undertaken, followed by steps designed to prevent *L. monocytogenes* from entering the processing plant. Both types of measures will reduce the contamination of the finished product. An awareness of the ubiquity of the bacterium, especially in those environments that favour its multiplication, has certainly resulted in improved control of manufacturing and retailing procedures. It is possible that the effectiveness of these measures together with warnings to susceptible individuals and the danger of consuming certain foods have markedly reduced the number of cases in the UK since 1989.

Finally, for those persons who fall into high-risk groups (e.g. pregnant women and the immunocompromised), it is advisable to avoid eating highly contaminated foods such as soft cheeses and pâté, and to re-heat cooked and chilled foods until piping hot.

4.2 *SHIGELLA* SPECIES

These are Gram-negative, facultatively anaerobic, non-spore-forming bacilli that do not ferment lactose and are non-motile.

4.2.1 Pathogenesis

There are two major mechanisms of pathogenesis, invasiveness and enterotoxin production.

1. Invasiveness. Shigellae proliferate in the gut lumen and are able to invade the mucosa of the terminal ileum and colon. Although they penetrate the epithelium and multiply within the cells, they rarely cause bacteraemia. However, because of the invasive process red blood cells may infiltrate into the lumen, resulting in characteristic bloody stools.
2. Enterotoxin production. Enterotoxin (also called Shiga toxin) activity typically induces intestinal tissue secretions, resulting in diarrhoea. This proteinaceous toxin is produced in large quantities only by *Sh. dysenteriae* serotype 1, although other *Shigella* spp. do produce small amounts. Since the toxin is proteinaceous, it is heat-labile and should be inactivated by adequate cooking. The toxin appears to bind to glycoproteins on the surface of a host cell, and inhibits protein synthesis by inactivating the 60S subunit of the ribosome, which results in cell death. Similar toxins, called Shiga-like toxins or 'verotoxins', are produced by certain strains of *Escherichia coli* and have already been described (section 3.6).

4.2.2 Clinical features

Shigella food poisoning (or shigellosis) may vary in severity from asymptomatic infection to fulminating dysentery; the severity of the symptoms depends a great deal on the species of shigella implicated in disease. When the causative organism is *Sh. dysenteriae*, there is often abdominal pain, fever, the frequent passage of bloody/fluid stools and the patient may also have headaches, nausea and undergo prostration. Alternatively, if the organism responsible is *Sh. sonnei*, there may only be a mild diarrhoea without fever. The incubation period may vary from 1–7 days, but is usually less than 4 days after ingestion of contaminated food.

4.2.3 Prognosis

Shigellosis is usually self-limiting and recovery often complete within 1–2 weeks after the onset of symptoms. However, very young, very old and/or debilitated patients with severe diarrhoea typically caused by *Sh. dysenteriae* may undergo rapid dehydration, which could prove fatal unless treated. Occasionally, antibiotic therapy is required; if possible it is advisable to determine the *in vitro* susceptibility of the organism first. This is because shigellae are known to acquire antibiotic resistance rapidly. This is often plasmid-mediated.

4.2.4 Incidence and epidemiology

Food poisoning outbreaks caused by shigellae have been reported worldwide; the incidence is greater in the USA, where there have been some large outbreaks, than the UK. For example, between 1986 and 1988 only two incidents were recorded in England and Wales and one of these was due to an imported *Sh. flexneri* infection. In contrast, 72 outbreaks were reported to the Center for Disease Control in the USA between 1961 and 1975. The majority of cases seen in the USA and the UK are due to *Shigella* species resident in the population, with *Sh. sonnei* being the most common and usually the least virulent. This is due to the fact that where foods are implicated in transmission, they will almost always have become contaminated by human faeces. A typical situation would arise where an infected person with poor personal hygiene handles cooked food. As with many other bacterial gastrointestinal infections, outbreaks usually occur during the summer months.

Cases of shigellosis due to *Sh. dysenteriae*, *Sh. flexneri* and *Sh. boydii* imported into the UK are usually responsible for more serious infections, which may be quite severe.

In the investigation of outbreaks, serotyping has been useful in discriminating among strains of *Sh. dysenteriae, Sh. flexneri* and *Sh. boydii*. However, as

Sh. sonnei has only one serotype and is commonly encountered, other typing systems have been developed including phage and bacteriocin typing. Bacteriocin or colicin typing is a form of typing in which strains of bacteria are distinguished on the basis of the bacteriocin(s) they produce or the bacteriocin(s) to which they are susceptible. More recently, plasmid profiles have been found to be useful in differentiating between strains.

4.2.5 Ecology

Shigella may be found in human faeces and in any other environment where faecal contamination has occurred, for example, water supplies affected by seepage of untreated sewage, or fomites, food, etc. which have suffered faecal soiling. Unlike salmonellae, *Shigella* spp. have rarely been isolated from animal species.

4.2.6 Foodstuffs

It has been difficult to draw conclusions from the limited number of cases in the UK, but the majority of recorded outbreaks in the USA have implicated salads which contained either potato, shrimp, tuna or chicken. In developing countries, where shigellae are commonly found in water supplies, any raw foods which have come in contact with water must be regarded as potential vehicles of infection.

4.2.7 Control

One of the most important factors that needs to be stressed is that *Shigella* spp. are highly infectious because they have a low infective dose: 100 bacteria or less can produce disease from contaminated fomites, food or water. This is obviously of relevance where personal hygiene is concerned, as only a minor lapse can have serious consequences. The best preventive measures would be good personal hygiene and health education for all people who handle food.

The majority of outbreaks of shigellosis worldwide are associated with drinking contaminated water; the disease may also be spread by person-to-person transmission by the faecal–oral route. Indirectly, therefore, it is of relevance to reduce the endogenous shigellae in the general population. This can be achieved more especially in developing countries by adequate chlorination of water supplies and the safe disposal of sewage.

4.3 *AEROMONAS* SPECIES

These are Gram-negative, facultatively anaerobic, non-spore-forming bacilli which are oxidase positive.

4.3.1 Pathogenesis

The pathogenicity of *Aeromonas* species (*A. hydrophila, A. sobria, A. caviae*) in human enteric infections is highly controversial. Although it is generally established that aeromonas (especially *A. caviae*) is a pathogen in children, this is not the case in adults. However, some studies have revealed aeromonas as the sole apparent pathogen in acute adult diarrhoea in a geographical area with a negligible asymptomatic carriage rate, thus lending support to the idea that it may be an enteropathogen in adults. Further recent evidence supporting the role of *A. hydrophila* in adult diarrhoeal illness was made when a cardiac surgeon developed symptoms, and before recognition of the *Aeromonas*-contaminated egg salad, the microbiologist investigating the surgeon's complaint ate some of the egg salad and also developed *A. hydrophila* diarrhoea. As recently as 1991, the first case of human intestinal aeromonas infection was reported, where the source was identified as contaminated shrimp.

A variety of potential virulence factors have been proposed, though definitive proof is lacking. These factors include:

1. heat-labile and/or heat-stable enterotoxins – a heat-labile enterotoxin has been found which cross-reacts with antitoxin produced against *V. cholerae* toxin and *E. coli* heat-labile enterotoxin;
2. cytotoxins, shown by their effect on HeLa cells;
3. haemolysins – one of which is called b-toxin and has properties similar to Kanagawa haemolysin (section 2.3);
4. adherence – especially to HEp-2 cells, which may be pilus-mediated.

Some studies have suggested no correlation between toxin production and the causation of gastroenteritis, but it may be that only strains with a specific combination of virulence factors (some of which may still be unknown) colonize the bowel and cause disease. Alternatively, host factors may play a major role.

4.3.2 Clinical features

Two types of enteric illness are associated with aeromonas. The most common is an illness similar to cholera, characterized by watery stools, i.e. diarrhoea and a mild fever. In young children vomiting may also occur. Less common is an illness similar to dysentery, characterized by diarrhoea and the presence of blood and mucus in stools.

Although no fully confirmed food poisoning outbreak has been described, a possible aeromonas-associated outbreak produced symptoms of nausea, vomiting, diarrhoea and stomach cramps. However, since the source of *A. hydrophila* infection in reported cases is generally unknown, it has not been possible to estimate the incubation period, until recently. However, it is now thought to be between 18 and 24 hours.

4.3.3 Prognosis

Patients with aeromonas-related enteric illness usually produce diarrhoea of 1–7 days' duration which is mild and self-limiting. However, there have been cases where symptoms have lasted for several weeks. Rarely, cholera-like symptoms are produced, when the patient will need rehydration and/or suitable antibiotic therapy.

4.3.4 Incidence and epidemiology

As a general rule, *A. hydrophila* has been associated with sporadic cases of food poisoning rather than with large outbreaks. However, although no large outbreaks have been described, it is possible that aeromonads were either not looked for or were not recognized in some reported outbreaks of apparently unknown aetiology. Other possibilities include a failure to examine either the patient's stools or implicated foods.

Both ribotyping and serotyping schemes are now available in the event of an outbreak to aid identification.

4.3.5 Ecology

Aeromonas species are aquatic bacteria that generally reside in fresh and brackish water. Usually, increasing numbers of aeromonas are found in aquatic environments in the summer, when the temperature of the water rises; in the winter months counts are generally low. These organisms can also be found in untreated sewage excreted from both symptomatic and asymptomatic persons and from farm animals.

4.3.6 Foodstuffs

Although raw shellfish have been implicated in two possible outbreaks, recent studies have shown *Aeromonas* species to be present in many foods, including raw meats, poultry and fresh vegetables; however, except in a shrimp-associated infection described above, it has not been established whether there was a direct link between the presence of these organisms in food and disease in humans.

4.3.7 Control

Two major problems associated with the control of *Aeromonas* species are first their frequent presence in food, and second, the fact that many strains are psychrotrophic. This means that they are capable of reaching high numbers, even in refrigerated foods. Fortunately, these are relatively heat-sensitive organisms and heat processing and/or cooking are usually effective in killing them. Since *Aeromonas* species appear to be ubiquitous in many types of foods, any foods that are eaten raw or undercooked are potentially hazardous. This is especially true of raw shellfish, where aeromonas may be actively concentrated inside molluscs; adequate cooking of shellfish is essential.

4.4 PLESIOMONAS SHIGELLOIDES

These are Gram-negative, facultatively anaerobic, non-spore-forming bacilli which are oxidase positive.

4.4.1 Pathogenesis

As in the case of *Aeromonas* species, there is some controversy regarding the enteropathogenicity of *P. shigelloides*. Like *Aeromonas* species, *P. shigelloides* has been implicated as an opportunist pathogen rather than as a cause of gastrointestinal disease in otherwise healthy persons. A further complication is that the organism has been found in up to 5% of stool specimens from persons without diarrhoeal symptoms. *P. shigelloides* is now considered by some to be invasive, and there have been some reports of enterotoxin activity; both a heat-labile and heat-stable toxin have been demonstrated. The present evidence suggests that the exact mechanism of enteropathogenicity is uncertain, and that more than one mechanism may be required to cause diarrhoea.

4.4.2 Clinical features

Few outbreaks of *P. shigelloides* food poisoning have been reported, but it would appear that the incubation period is usually 1 day, although it may sometimes extend to 2 days. The most common symptoms produced are diarrhoea, abdominal pain and nausea, with fever, headache and vomiting being less common.

4.4.3 Prognosis

This type of gastroenteritis is usually mild and self-limiting and the duration of the illness is between 1 and 7 days in previous healthy individuals. Rehydration and/or antibiotic treatment are rarely necessary. Infection in an immunocompromised patient may well be much more severe, with concomitant alterations in therapy required.

4.4.4 Incidence and epidemiology

There have been few reports of food-related cases of *P. shigelloides* infection worldwide, even in developed countries such as the USA and the UK where isolation and identification of the organism are likely to be more successful than in developing countries. For example, in England and Wales there were only four sporadic cases between 1986 and 1988. However, there does appear to be a correlation between reported cases and the consumption of foods of aquatic origin. This is related to the fact that *P. shigelloides* occurs naturally in water. In the rare event of an outbreak, biotyping and/or serotyping may prove to be useful.

4.4.5 Ecology and foodstuffs

Like *Aeromonas* species, *P. shigelloides* are aquatic bacteria that generally reside in fresh and brackish water, though they can also survive in conditions of comparatively high salinity. In addition to humans, other sources of *P. shigelloides* include fish, reptiles, crustaceans, birds and other mammals. It has been suggested that freshwater fish are one of the primary reservoirs of *P. shigelloides,* which may help explain why the organism is ubiquitous in fresh water. Another similarity with *Aeromonas* species is that *P. shigelloides* are usually found with increased frequency during the summer months.

Very few foods have been implicated with any degree of certainty in *P. shigelloides* food poisoning; those that have include fish, crabs and oysters.

4.4.6 Control

Given our knowledge of the ecology of *P. shigelloides*, we must assume that water is contaminated unless it is known to be properly treated. Adequate cooking of natural water-related foods must, therefore, be ensured, and the necessary precautions should be taken to prohibit water contamination of cooked/processed food. All strains of *P. shigelloides* are heat-sensitive and are readily killed by heating at 60°C for 30 minutes. Although these features are not common to all strains, some will grow at pH 4.0, while others will grow at 8°C. This latter fact has obvious implications for food processing and refrigeration. It is possible that high numbers of certain strains of these organisms could be achieved, even in refrigerated foods. Finally, as *P. shigelloides* may be actively concentrated within oysters, it is essential that these shellfish are adequately cooked.

4.5 *STREPTOCOCCUS PYOGENES* AND *ENTEROCOCCUS FAECALIS*

These are Gram-positive, facultatively anaerobic cocci that are catalase negative (Table 4.6).

Table 4.6 Differentiation of *Streptococcus pyogenes* and *Enterococcus faecalis*

Property	St. pyogenes	Ent. faecalis
Haemolysis	ß	$-/\beta/\alpha$
Polysaccharide (Lancefield) antigen	A	D
Growth at 45°C	–	+
Growth at pH 9.6	–	+
Growth on 10% and/or 40% bile	–	+
Survival at 60°C for 30 minutes	–	+

4.5.1 Pathogenesis

Traditionally, *Streptococcus pyogenes*, also known as the Lancefield group A streptococcus, is associated with many human diseases, most commonly pharyngitis and/or tonsillitis and late complications of infection such as rheumatic fever and acute glomerulonephritis. *Enterococcus faecalis* (previously called *Streptococcus faecalis*), which is one of the Lancefield group D streptococci, is implicated in urinary tract and wound infections, intra-abdominal abscesses and endocarditis. As enterococci are resistant to most antibiotics, the introduction of new broad-spectrum agents has allowed these organisms to proliferate and cause serious superinfections, especially in immunocompromised patients. However, other β-haemolytic streptococci, *St. pyogenes* and the recently re-named *Ent. faecalis*, have all been implicated in food poisoning to a significant degree, causing classical symptoms such as diarrhoea, as well as being responsible for other food-borne infections such as epidemic tonsillitis (*St. pyogenes* only). Unfortunately, no enterotoxin or other mechanisms of enteropathogenicity have been found in streptococcal strains associated with food poisoning, and therefore we can only state that in such cases these organisms were presumed, but not proven, to be the cause of gastroenteritis. In some instances non-haemolytic or α-haemolytic strains of *Ent. faecalis* have been implicated.

4.5.2 Clinical features

Although few outbreaks have been reported, the incubation period is generally short, ranging from a few hours to 48 hours. Symptoms have been described as milder than those associated with *Staphylococcus aureus* food poisoning, with diarrhoea as the prominent feature; nausea and vomiting are seen less frequently.

4.5.3 Prognosis

Recovery is usually rapid, often within 24 hours of the onset of symptoms. However, rarely, in elderly and immunocompromised patients severe diarrhoea and hypotension may prove fatal, and in these cases appropriate antibiotics should be administered together with other forms of therapy.

4.5.4 Incidence and epidemiology

Ent. faecalis has been associated with a larger number of outbreaks of gastro-enteritis and over a longer period of time (since 1926) than has *St. pyogenes*. However, the evidence for disease causation has consisted simply of the detection and the presence of the organisms in large numbers in suspected foods, rather than the identification of specific mechanisms of en-teropathogenicity. In relative terms, the number of recognized cases and outbreaks of gastroenteritis due to *Ent. faecalis* is low in the USA and Western Europe. It should of course be realized that in routine investigations of gastroenteritis, faeces are not usually examined with a view to isolating either *St. pyogenes* or *Ent. faecalis*. Even if this was the case, it would still be extremely difficult to differentiate between commensal enterococci (normal flora) and suspected pathogenic enterococci. Therefore, it is even more difficult to assess the situation worldwide, as the problems of laboratory diagnosis would be even greater in developing countries. Gastroenteritis due to *St. pyogenes* is rare, but has been fatal and is thought to be dose-dependent.

4.5.5 Ecology and foodstuffs

St. pyogenes is found commonly on human skin and mucous membranes, especially those of the respiratory tract, and may also colonize the rectum. As the name suggests, *Ent. faecalis* is a normal inhabitant of the gastro-intestinal tract.

Although a number of different foods have been implicated in the causation of gastroenteritis, dairy products, eggs, meats and salads occur most frequently in reports.

4.5.6 Control

Since *Ent. faecalis* is present in the gastrointestinal tract, personal hygiene is of the utmost importance for all food-handlers. If this organism is allowed to contaminate food, there are a number of factors which make control difficult. These include greater heat resistance (i.e. survival after heating at 60°C for 30 minutes) and growth over a wider temperature range (10–45°C) than many other bacteria (some strains may also grow at 4°C). Foods should, therefore, be cooked thoroughly at a high enough temperature, and then cooled rapidly and stored under adequate refrigeration (4°C).

Good personal hygiene would also be of importance in helping to eliminate infections due to *St. pyogenes*, in addition to the obvious precautions of excluding food-handlers who have skin lesions or present with a classical respiratory illness such as a sore throat.

5 Mycotoxic fungi

M.O. Moss

5.1 INTRODUCTION

The fungi are a diverse group of organisms which have in common a eukaryotic structure, a heterotrophic metabolism and an outer wall. Thus, unlike plants, fungi require organic carbon compounds, although the presence of a wall means that, unlike animals, they have to feed by absorption of soluble nutrients rather than by ingestion of particulate food and subsequent digestion. Many fungi can metabolize complex insoluble materials, such as lignocellulose, but these materials have to be degraded by the secretion of appropriate enzymes outside the wall. A number of fungi are parasitic on animals, plants and other fungi, and some of these parasitic associations have become very complex and may even be obligate. The fungi are justifiably considered to be a Kingdom of living organisms and, within this Kingdom, the food mycologist is primarily concerned with those fungi conveniently referred to as yeasts on the one hand and as moulds, mushrooms and toadstools (known collectively as filamentous fungi) on the other. Of these, we shall be concerned principally with the moulds as producers of toxins in foods which have been associated with a range of human diseases, from gastroenteritis to cancer.

The yeasts are essentially single-celled fungi which, although morphologically simple, are a highly specialized group associated with such nutrient-rich environments as the nectaries of plants and the body fluids of insects and other animals. A few species of yeasts, such as *Candida albicans* and *Cryptococcus neoformans*, are potentially pathogenic in humans. Although many species are associated with food spoilage, and several are used in food production, they are not known to produce toxic metabolites in association with food, unless one considers ethanol and its simple derivatives, such as esters, as toxic compounds! In contrast, the filamentous fungi are able to synthesize a diverse range of low-molecular weight compounds (many of which are species-specific or even strain-specific), which are collectively known as secondary metabolites. This is not the place to consider the physiology of secondary metabolite biosynthesis, but it should be recognized that these compounds are chemically very diverse and this is reflected in the wide range of properties with which they are associated. Thus many are coloured, many have antibiotic properties, many are phytotoxic (toxic to plants) and a significant number are toxic to animals including mammals such as humans.

The filamentous fungi grow over and through their substrate by processes of hyphal tip extension, branching and anastomosis. This leads to the production of an extensive mycelium, which absorbs nutrients and also secretes secondary metabolites, as well as enzymes to degrade macromolecules. Some species have been especially successful in growing at relatively low water activities (Chapters 9 and 11), which allows them to colonize commodities such as cereals which would otherwise be too dry for the growth of many micro-organisms. Frequently, when moulds attack foods, they do not cause the kind of putrefactive breakdown associated with some bacteria, and their presence is often tolerated. Indeed, some of the changes brought about by the growth of certain fungi on a food may be desirable, and there is a strong tradition of products such as mould-ripened cheeses, mould-ripened sausages, and food produced by deliberately growing moulds on plant products such as soyabeans (e.g. tempeh, miso and soy sauce). However, what concerns us here is the ability of some fungi to produce toxic metabolites which may cause harm when eaten.

5.1.1 Mycotoxins and mycophagy

The vegetative structure of the filamentous fungi is essentially based on the growth form of the spreading, branching, anastomosing mycelium, and has a relatively limited morphological diversity: to the inexperienced eye the mycelium of one species looks like the mycelium of any other. It is in the structures associated with spore production and dispersal that the developmental and morphological diversity of the filamentous fungi become truly apparent. Many are microscopic, and such fungi are conveniently referred to as the moulds, but among the basidiomycetes and ascomycetes there are species producing prodigiously macroscopic fruit bodies, the mushrooms and toadstools. They have evolved as very effective structures for the dispersal of spores produced as a result of sexual reproduction. These two aspects of fungal morphology have led to two distinct branches in the study of fungal toxins.

The mycotoxins are a subgroup of the secondary metabolites of moulds which may contaminate food or animal feeds, or the raw materials for their manufacture, and which happen to be toxic to humans and domestic animals. Their study, and the legislation associated with their control, are based on their being considered as adulterants of foods or animal feeds.

On the other hand, the macroscopic fruiting bodies of some ascomycetes and many basidiomycetes have provided a traditional source of food in many parts of the world for many thousands of years. Although many different species are edible, only a few lend themselves to large-scale cultivation; in the UK production of the cultivated mushrooms, *Agaricus bisporus*, is a major multi-million pound industry. There are, however, many more species which are collected from the wild and enjoyed as food. This group of fungi includes a number of species which produce toxic metabolites in their fruit bodies but, because these toxins are a natural constituent of fruiting bodies deliberately ingested (usually as a result of mistaken identify), they are not considered as mycotoxins. This is a somewhat arbitrary distinction, based on human behaviour and not on the chemistry, biochemistry or toxicology of the compounds. Table 5.1 lists a few of the more important toxins associated with macroscopic fungal fruiting bodies.

Fortunately, there are relatively few species of agarics which can be considered as deadly poisonous, but they do include the deathcap, *Amanita phalloides*, a quarter of a cap of which can be lethal to a healthy adult, and species of *Cortinarius* which are still foolishly mistaken for edible wild fungi. In both these cases the toxins cause irreversible damage to the liver and kidneys.

Before considering the mycotoxigenic moulds in detail, it is worth emphasizing a major difference between the toxic metabolites of fungi and the toxins of most of the bacteria associated with food poisoning. The former are relatively low molecular weight compounds, although their chemistry may be very complex, while the latter are macromolecules such as polypeptides, proteins or lipopolysaccharides. Some of these prokaryote toxins, such as botulinum toxin, may be a million times more toxic than even the most virulent mycotoxin! One interesting exception to this generalization is an unconventional bacterial food poisoning associated with a traditional food produced in parts of Indonesia. A form of tempeh made from coconut flesh is produced by allowing moulds such as *Rhizopus* and *Mucor* to grow over the coconut. Occasionally the process becomes contaminated with a bacterium

Table 5.1 Toxic compounds of some 'toadstools'

Toxin	Species	Toxic effects
Coprine	*Coprinus atramentarius*	Considerable discomfort when consumed with alcohol; reaction occurs within 10–15 minutes to several hours.
Illudin	*Omphalotus olearius*	Gastrointestinal irritation, vomiting; symptoms occur within 0.5–2 hours.
Amatoxin	*Amanita phalloides*	Liver and kidney damage, death unless treated; symptoms develop within 10–14 hours and death may occur from 4–7 days.
Orellanin	*Cortinarius orellanus*	Irreversible kidney damage, death or very slow recovery; vomiting may occur within 4–9 hours but signs of kidney damage to not appear for 3–14 days.
Psilocybin	*Psilocybe cubensis*	Hallucinogenic; symptoms within 30–60 minutes.

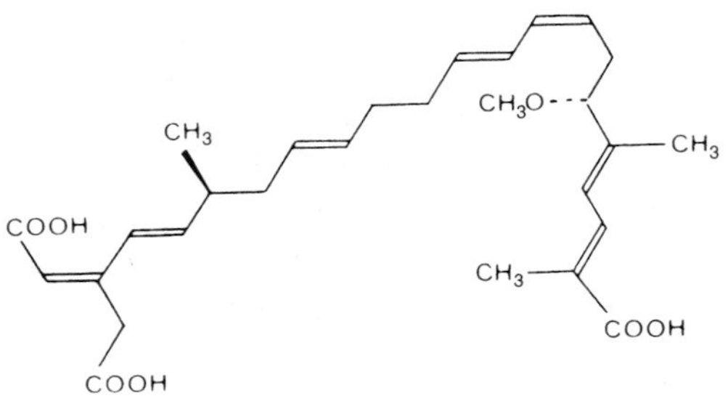

Bonkrekic acid

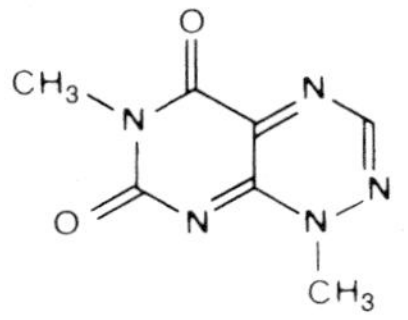

Toxoflavin

Figure 5.1   Toxic metabolites of *Pseudomonas cocovenenans.*

referred to as *Pseudomonas cocovenenans*, which is known to produce at least two low molecular weight toxic metabolites, bonkrekic acid and toxoflavin (Figure 5.1). Consumption of this contaminated material can lead to severe illness and even death.

The examples of mycotoxicoses described below will generally be associated with particular species of mould, but it should be appreciated that some species produce several toxic metabolites, and some individual mycotoxins may be produced by several relatively unrelated species of mould.

Although many species from a large number of unrelated genera of fungi can produce toxic metabolites, and a significant number of these have been implicated in poisoning, especially of farm animals, there are three genera which stand out as being especially important in the context of well-being of people, namely *Fusarium, Penicillium* and *Aspergillus*.

5.2 *FUSARIUM* SPECIES

The genus *Fusarium* is very important because of the involvement of some species with economically devastating disease of crop plants causing wilts, blights, root rots and cankers. Members of the genus are also agents of biodegradation, causing the post-harvest spoilage of crops in storage. A few species, such as *Fusarium solani*, are known to be associated with infections in humans in which *Fusarium* can be an opportunist pathogen. On the other hand, another species of *Fusarium* is grown on an industrial scale in the production of mycoprotein for incorporation into a wide range of interesting food products such as Quorn™. Species of *Fusarium* are also associated with the production of a large number of chemically diverse metabolites, some of which are mycotoxins.

5.2.1 *Fusarium sporotrichioides* and *Fusarium poae*

Outbreaks of a serious disease known as alimentary toxic aleukia, which is also referred to as septic angina and acute myelotoxicosis, occurred during famine conditions in a large area of Russia from 50–60° latitude and between 40–140° longitude. A particularly severe outbreak was reported during the period 1942–1947, but there have been reports of the disease in Russia since the 19th century.

Although intensive studies in Russia itself demonstrated that the disease was associated with the consumption of cereals moulded by *F. sporotrichoides* and *F. poae*, the nature of the toxin was made clear as a result of studies of a veterinary problem in the USA. It was shown that severe cases of dermonecrosis in cattle had been caused by a *Fusarium* metabolite, referred to as T-2 toxin, which is one of the most acutely toxic of a family of compounds called trichothecenes

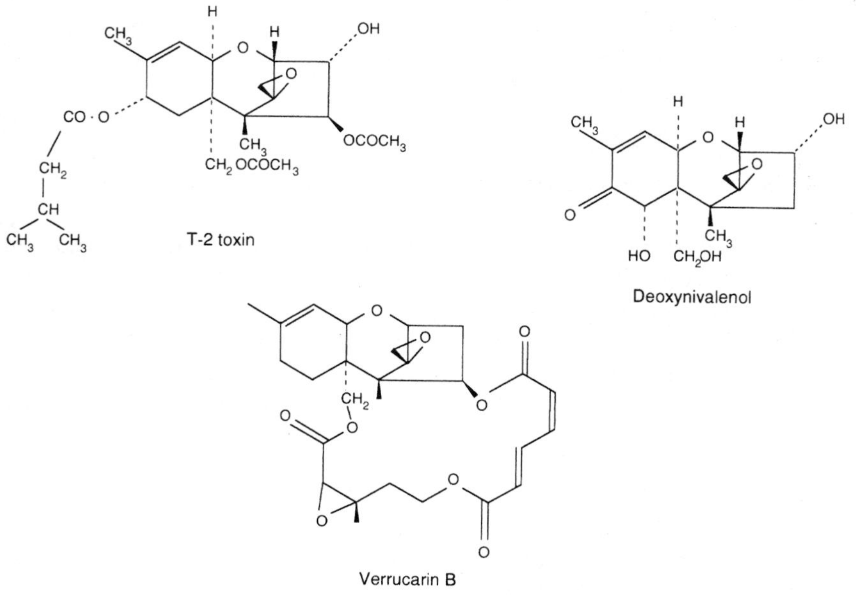

Figure 5.2 Examples of trichothecene mycotoxins.

(Figure 5.2). There is good evidence that T-2 toxin was also a major agent in the development of alimentary toxic aleukia in humans.

Following consumption of cereals which have become moulded after exposure to cold, wet conditions, the first signs of discomfort appear very rapidly and may last for 3–9 days; these are associated with damage of the mucosal membrane systems of the mouth, throat and stomach, followed by inflammation of the intestinal mucosa. Symptoms such as bleeding, vomiting and diarrhoea, which are all associated with damage of mucosal membrane systems, are common, but recovery at this stage is relatively easy if the patient is given a healthy, uncontaminated diet. Continued exposure to the toxin leads to damage of the bone marrow and the haematopoietic system, followed by anaemia and a decrease in erythrocyte and platelet counts. This phase might last from 2–8 weeks and is followed very rapidly by the final stages. Further inflammatory degeneration usually occurs, associated with weakening of capillary walls and subsequent haemorrhaging. Necrotic tissue frequently becomes infected with bacteria, and many deaths undoubtedly occur as a result of secondary bacterial and viral infections. As well as giving rise to this sequence of acute symptoms, the trichothecenes are known to be immunosuppressive, and this undoubtedly contributed to the sensitivity of those people already suffering from the acute toxicity to relatively trivial infectious agents. Unlike aflatoxin, the acute toxicity of T-2 toxin is remarkably uniform over a wide range of animal species (Table 5.2), and it is reasonable to assume that the LD_{50} for humans will be in the same range. Improved harvesting and storage have hopefully eliminated alimentary toxic aleukia from Russia, although similar cases of poisoning following consumption of *Fusarium* toxins may occur in other parts of the world ravaged by war and famine.

5.2.2 *Fusarium graminearum* and related species

In Japan, there have been many records of nausea, vomiting and diarrhoea in the past associated with the consumption of wheat, barley, oats, rye and rice contaminated by species of *Fusarium*. Because of the discolouration of the foods caused by the fungi, they were frequently referred to as red-mould disease. Outbreaks of poisoning in humans were reported in 1946 and 1955 as well as during earlier periods, and the mould most frequently incriminated was

Table 5.2   Some LD_{50} values for T-2 toxin

Species	LD_{50} (mg/kg)
Guinea pig	3.1 (oral)
Seven-day-old chick	4.0 (oral)
Mouse	5.2 (intraperitoneal)
Rat	5.2 (oral)
Trout	6.1 (oral)

F. graminearum. During the earlier studies, isolates of fusaria had been mis-identified as *Fusarium nivale* and the trichothecene toxins isolated from them were called nivalenol and deoxynivalenol. It is now known that *F. nivale*, which may not even be a species of *Fusarium*, does not produce trichothecenes at all but the names of the toxins have been retained!

Deoxynivalenol was also isolated and characterized as the vomiting factor and possible feed refusal factor in an outbreak of poisoning of pigs fed on moulded cereals in the USA. In this context it was given the rather descriptive name of vomitoxin, but is now most widely known as deoxynivalenol or DON. Although it is a trichothecene, deoxynivalenol is much less acutely toxic than T-2 toxin, having an LD_{50} of 70 mg/kg in the mouse. Nevertheless, it is considerably more widespread than T-2 toxin and has caused some concern in the production of winter wheat and winter barley in such countries as Canada. In 1980 there was a 30–70% reduction in the yields of spring wheat harvested in the Atlantic provinces of Canada due to infections with *F. graminearum* and *F. culmorum*, both of which may produce DON and zearalenone. It is not clear whether DON and other trichothecenes are as immunosuppressive as T-2 toxin, but it seems prudent to reduce exposure to a minimum.

Totally unrelated to the trichothecenes, and first identified as the agent responsible for vulvovaginitis in pigs fed on mouldy maize, zearalenone is recognized as an oestrogenic mycotoxin. Its acute toxicity is minimal, but it is very common in cereals such as maize, wheat and barley, and there should be concern about the long-term exposure of the human population to low levels of such an oestrogen. A number of instances of an imbalance in the development of secondary sexual features among children consuming maize products con-taminated with zearalenone have been discussed among mycotoxicologists, but the author is not aware of any literature reports. The pig remains one of the animals most sensitive to the oestrogenic effects of this toxin. Zearalenone (Figure 5.3) is so named because of its isolation from the ascomycete *Gibberella zeae*, which has an asexual, or imperfect stage known to mycologists as *F. graminearum*. It is the latter form which is most commonly isolated from cereals.

Zearalenone, and the corresponding alcohol zearalenol, are known to have anabolic properties, and there is probably as large a patent literature based on

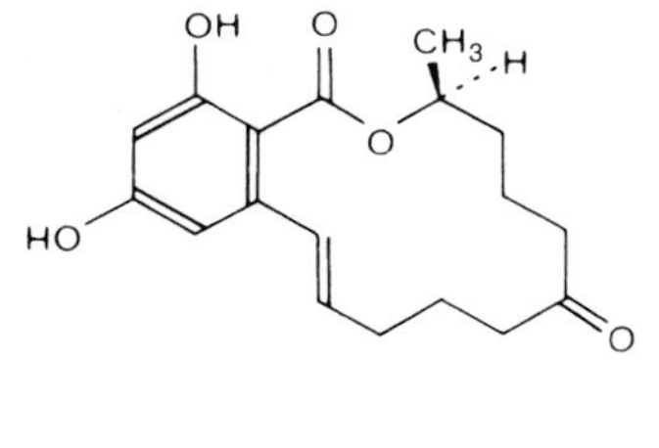

Zearalenone

Figure 5.3 Zearalenone.

this biological activity as there is information about the metabolites as mycotoxins. Although its use as a growth-promoting agent for farm animals is banned in some countries, it is permitted in others. This difference can lead to problems in international trade, because detectable levels of zearalenone may be identified in the meat of animals fed on diets contaminated with it.

5.2.3 *Fusarium moniliforme*

In some parts of northern China and in the Transkei in southern Africa there are regions of high incidence of oesophageal cancer in humans, and the epidemiology of the disease would seem to implicate the consumption of moulded cereals and involve a role for mycotoxins. A species of *Fusarium* belonging to a different group from those already considered, *F. moniliforme*, would seem to be the fungus most likely to be involved. Strains of this species are associated with a disease of rice which has been a particular problem in China, and other, probably distinct, strains are commonly isolated from maize grown in southern Africa. *F. moniliforme*, and the related species *Fusarium subglutinans*, occur in the tropics and subtropics, and there has unfortunately been some confusion over their identification. These species do not produce trichothecenes but are nevertheless very toxigenic, and their presence on animal feeds is associated with field outbreaks of equine leukoencephalomalacia and laboratory cases of liver cancer in rats. *F. moniliforme* is especially common on maize in tropical Africa, and the incorporation of such contaminated maize into human food is strongly correlated with the occurrence of oesophageal cancer in humans in some parts of the Transkei. One of the first mycotoxins isolated and characterized was called moniliformin because it was presumed to have been produced by *F. moniliforme*. It is now considered that moniliformin is actually produced by strains of *F. subglutinans* and not *F. moniliforme*. Nevertheless, it is the latter species which is especially associated with oesophageal cancer in humans. A number of complex metabolites have been isolated and characterized from cultures confirmed to be of *F. moniliforme* (Figure 5.4). They include fusarin C, which is mutagenic, and the fumonisins, which are the most likely candidates to account for the carcinogenic activity of this species. There is now no doubt that the fumonisins cause leukoencephalomalacia in horses, pulmonary oedema syndrome in pigs, and hepatic cancer in rats. However, caution is required about extrapolating laboratory tests which relate carcinogenic activity to a disease in humans; it is probable that there is still more to learn about this remarkably important mould.

5.3 *PENICILLIUM* SPECIES

Penicillium is much more common as a food spoilage mould in Europe than *Aspergillus*, with species such as *Penicillium italicum* and *Penicillium digitatum*

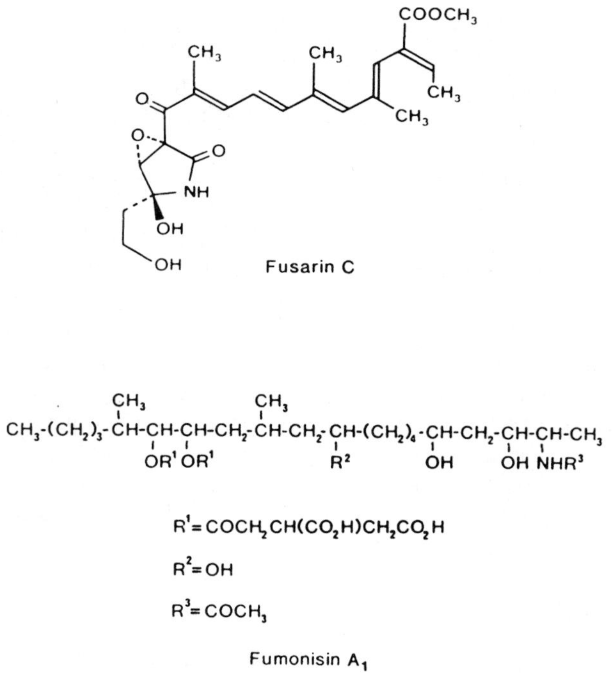

Fusarin C

$R^1 = COCH_2CH(CO_2H)CH_2CO_2H$

$R^2 = OH$

$R^3 = COCH_3$

Fumonisin A$_1$

Figure 5.4   Toxic metabolites of *Fusarium moniliforme*.

causing blue and green mould respectively of citrus fruits, *Penicillium expansum* causing a soft rot of apples, and a number of other species associated with the moulding of jams, bread and cakes. Several species have a long association with mould-ripened foods, such as *Penicillium roquefortii*, which is used in all the blue cheeses of Europe, and *Penicillium camembertii*, used in the mould-ripened soft cheeses.

5.3.1 *Penicillium expansum*

A mycotoxin which is especially associated with *P. expansum* and is of medical interest is patulin, which was first described in 1942 as a potentially useful antibiotic with a wide spectrum of antimicrobial activity. Indeed, it was discovered several times during screening programmes for novel antibiotics, and this is reflected in the many names by which the compound was known, including claviformin, clavicin, expansin, penicidin, mycoin, leucopin, tercinin and clavatin. At one time there was some hope that patulin might have had some use in the treatment of the common cold; however, it was soon rejected as an antibiotic of any clinical significance because it turned out to be too toxic in animal trials. There must be a considerable amount of information about toxic mould metabolites associated with the descriptions of unsuccessful antibiotics, which may not be available for consideration in the context of mycotoxin studies. It was not until 1959 that an outbreak of poisoning of cattle being fed on an emergency ration of germinated barley malt sprouts alerted the veterinary profession to patulin as a mycotoxin. In this instance the producing organism was *Aspergillus clavatus*, but the same toxin has been implicated in several outbreaks of poisoning from such diverse materials as apples infected with *P. expansum* to badly stored silage infected with a species of *Byssochlamys*. As far as humans are concerned, patulin is not known to have caused any overt illness, but the common association of *P. expansum* with apples, and the increasing consumption of fresh apple juice as a beverage, has caused some concern and has led to the routine analysis for patulin in fruit juices. The third report (1993) on mycotoxins of the UK Ministry of Agriculture, Fisheries and Food Steering Group on Chemical Aspects of Food Surveillance provided details of a survey of clear and cloudy retail apple juices made in 1992. Of 32 samples analysed, five had levels of patulin above 50 µg/kg, a level which has been set as an advisory level to be achieved by good manufacturing practice and the exclusion of mouldy fruit from production. Patulin is not a particularly stable metabolite, but is most stable at the relatively low pH values of apple juices, although it is destroyed during the fermentation of apple juice to cider. Although the toxicological significance of this metabolite to humans is a matter for speculation, the presence of patulin in fresh apple juice is a useful indicator of the use of poor-quality fruit and possibly poor hygiene in its manufacture.

5.3.2 *Penicillium citreo-viride* and *Penicillium islandicum*

A very complex range of disorders has been described in Japan since the end of the last century, associated with the presence of several toxigenic species of *Penicillium* growing on rice. This moulded rice is usually discoloured yellow, and several of the toxic metabolites implicated are themselves yellow pigments. There was an early awareness that moulds might be responsible for cardiac beri-beri, and in 1938 it was demonstrated that *P. citreo-viride* (*P. toxicarium*) and its metabolite citreoviridin might be responsible. The occurrence of liver tumours in humans associated with the consumption of moulded rice was recognized in Japan shortly after the Second World War, and one of the most toxic species of *Penicillium* studied by Japanese mycotoxicologists was *P. islandicum*. This species can produce both luteoskyrin and islanditoxin (Figure 5.5). The former is not very acutely toxic (oral LD_{50} in mice, 221 mg/kg body weight) but it has been shown to be carcinogenic in experimental animals. The latter is very much more acutely toxic (oral LD_{50} in mice, 6.6 mg/kg body weight) and can cause severe liver damage leading to rapid death. A mixture of the two could well produce complex symptoms and, if the nephrotoxin citrinin produced by *Penicillium citrinum* and citreoviridin were also present in mouldy rice, then it is not surprising that 'yellow rice disease' was a difficult and variable food poisoning to understand.

5.3.3 *Penicillium verrucosum*

Although ochratoxin A, which is a potent nephrotoxin, was first isolated from *Aspergillus ochraceus* in South Africa, it has been most extensively studied as a contaminant of cereals such as barley infected with *P. verrucosum* in temperate countries, such as those of Northern Europe. This is because it is known to be a major aetiological agent in kidney disease in pigs, and may be passed through the food chain in meat products to humans. Ochratoxin is relatively stable and has been detected unchanged in the serum of a number of people during a recent study carried out in Europe, thus demonstrating that it certainly occurred in their food.

In humans there is a disease known as Balkan endemic nephropathy, the epidemiology of which is still a mystery. This debilitating disease may be associated with the presence of low levels of nephrotoxic mycotoxins, such as ochratoxin, in the diet of people who traditionally store mould-ripened hams

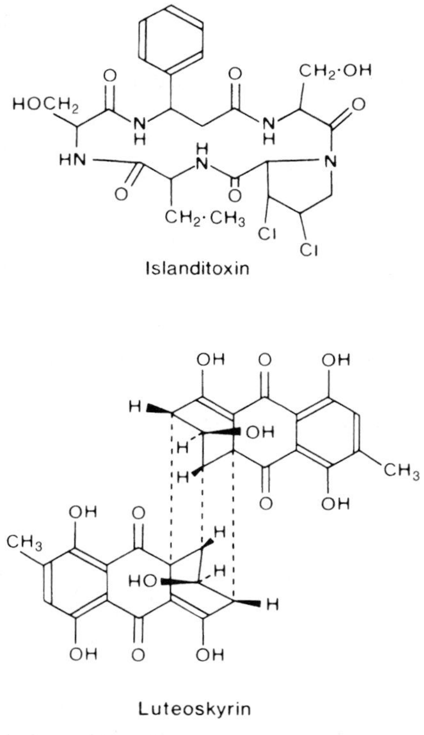

Islanditoxin

Luteoskyrin

Figure 5.5 Toxic metabolites of *Penicillium islandicum*.

for long periods of time. However, it does seem probable that if ochratoxin itself was implicated this would have been clearly established by now. The search is continuing in the laboratories of Imperial College, London, and nephrotoxic metabolites quite different from ochratoxin have been extracted from isolates of *Penicillium aurantiogriseum* and *Penicillium commune*.

Like patulin, the nephrotoxin metabolite citrinin, produced by *P. citrinum*, was also first discovered as a potentially useful antibiotic, but again rejected because of its toxicity. It is probably not as important as ochratoxin although, as mentioned above, it may be implicated in the complex epidemiology of 'yellow rice disease'.

5.4 *ASPERGILLUS* SPECIES

In 1959 a very singular event occurred, which initiated the international interest which now exists in one group of mycotoxins, the aflatoxins. The event was the deaths of several thousand turkey poults and other poultry on farms in East Anglia. Because of the implications for two major industries, the turkey industry and the manufacture of pelleted feed which supported it, a considerable effort was put into understanding the aetiology of this outbreak of what was initially referred to as turkey X disease. Although the name implies that they suffered from an infectious disease, it was soon demonstrated that the birds had been poisoned by a contaminant in the groundnut meal used as a protein supplement in the pelleted feed. This contaminant, which fluoresced intensely under ultraviolet light, was shown to be an acutely hepatotoxic form of aflatoxin, produced by the mould *Aspergillus flavus* growing on the groundnuts used for the production of groundnut meal. A few years later chronic toxicity tests on rats showed that, for this species, aflatoxin is among the most carcinogenic compounds known. This demonstration made it possible to rationalize the aetiology of diseases such as liver carcinoma in rainbow trout and hepatitis X in dogs, which had been described nearly a decade earlier but had remained a mystery. It also led to the establishment of very sensitive analytical methods for aflatoxins and the realization that their occurrence was widespread in commodities such as groundnuts and maize, much of which might be destined for human consumption.

5.4.1 The aflatoxins

Aflatoxins are known to be produced by two closely-related species of mould, *A. flavus* and *Aspergillus parasiticus,* both of which are especially common in the tropics and subtropics. More recently a third species, *Aspergillus nomius*, has been recognized as aflatoxigenic, but the frequent reports in the early literature of the production of aflatoxins by other species even belonging to different genera, are usually the result of mistakes.

Initially, it was considered that aflatoxin contamination was essentially a problem of poor storage of commodities after harvest, allowing the growth of storage fungi such as aspergilli and penicillia, with consequent formation of mycotoxins, including aflatoxin. Indeed, these conditions of high humidity and warm temperature can give rise to the highest levels of aflatoxin in food, often exceeding the upper limit initially established by the Food and Agricultural Organization (FAO) and the World Health Organization (WHO) of 30 μg/kg in foods for human consumption. These agencies faced a different dilemma when setting these limits, and this is reflected in the observation that 'clearly the group would have preferred a lower figure, but felt that the danger of malnutrition was greater than the danger that aflatoxin would produce liver cancer in man'. Meanwhile, many developed countries had set legislative or guideline levels that were more stringent than the limits set initially by the FAO/WHO. Some of the more recent maximum tolerance levels for aflatoxins are shown in Table 5.3.

Like many microbial secondary metabolites, the aflatoxins are a family of closely related compounds, the most toxic of which is referred to as aflatoxin B_1. Aflatoxin B_1 (Figure 5.6) is an acute toxin as well as being carcinogenic and immunosuppressive but, in contrast to T-2 toxin, the precise nature of the response to the consumption of aflatoxin is dependent on species, sex and age. Some animals, such as the day-old duckling and the adult dog, are remarkably sensitive to the acute toxicity of aflatoxin B_1, with LD_{50} values of 0.35 and 0.5 mg/kg body weight respectively, while others, such as the adult rat and the mouse, are more resistant (LD_{50} ca. 9 mg/kg). Not all animals respond to the carcinogenic activity of aflatoxin, but for the rat and rainbow trout aflatoxin B1 is one of the most carcinogenic compounds known. Indeed, as little as 4 μg in the diet of rainbow trout gave rise to a 25% incidence of hepatoma within 12 months. What about humans? Are people as sensitive as the dog or as resistant as the rat to the acute toxicity? Does aflatoxin cause liver cancer in humans?

There have been a number of isolated cases of deaths attributed to aflatoxin in food, but a particularly tragic demonstration of the acute toxicity of aflatoxin

Table 5.3 Maximum tolerated levels for aflatoxin in foodstuffs

Country	Commodity	Tolerance (μg/kg)
Australia	Peanut products	15
Belgium	All foods	5
Canada	Nuts and nut products	15
China	Rice and other cereals	50
France	All foods	10
	Infant foods	5
United Kingdom	Nuts and nut products, dried figs	4*
United States	All foods	20

*Statutory Instrument [1992] No. 3236

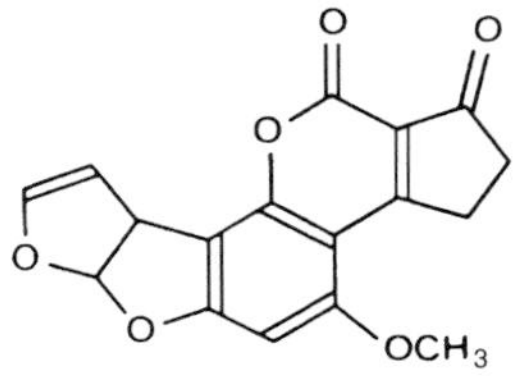

Aflatoxin B₁

Figure 5.6   The structure of aflatoxin B₁.

to humans was reported in India in 1974, when a large outbreak of poisoning occurred, involving nearly 1000 people, of whom nearly 100 died. From the concentrations of aflatoxins analysed in the incriminated mouldy maize it is possible to extrapolate an estimate of the LD_{50} for aflatoxin B_1 in humans as very approximately 5 mg/kg body weight, thus lying somewhere between that for the dog and the rat.

Although aflatoxin may be considered among the most carcinogenic of natural products for some animals, it is still not clear whether it is a carcinogen for humans. The pathogenesis of liver cancer in parts of the world such as the African continent is undoubtedly complex, and the initial demonstration of a correlation between exposure to aflatoxin in the diet and the incidence of liver cancer has to be considered with caution. Many animal studies have shown that the female is more resistant to all aspects of the toxicity of aflatoxin than the male, and these correlations have even provided some evidence that suggests the same would be true of men and women. However, it is also known that an even stronger correlation occurs between the presence of hepatitis B virus and primary liver cancer in humans, and there is an increasing consensus that these two agents act synergistically. The need to keep an open mind is further demonstrated by reports in 1987 of studies from the former Czeckoslovakia of 38 patients with primary liver cancer. Viral hepatitis antigen was detected in the sera of eight of the patients, while aflatoxin B1 was detected, using a radioimmunoassay, in the livers of 27 of 34 of the patients studied. Although liver cancer may be attributable to exposure to aflatoxin in parts of Africa, it is necessary to ask why it is not also more prevalent in India, where dietary exposure to aflatoxin also occurs; genetic differences may play a role here. In India cirrhosis of the liver is more common. There is still a lot to learn about the role of aflatoxin in liver disease in different parts of the world.

When one sees a diverse range of responses to the toxic effects of a compound it is a reasonable assumption that the compound is metabolized in the animal body, and that the resulting toxicity is the final result of this metabolic activity. This is certainly the case with aflatoxin B_1, from which a very wide range of metabolites is formed in the livers of different animal species. Thus the cow is able to hydroxylate the molecule and secrete the resulting aflatoxin M_1 (Figure

5.7) in the milk, hence affording a route for the contamination of milk and milk products in human foods, even though these products have not been moulded.

Two other important metabolites which are almost certainly implicated in the toxicity of aflatoxin B_1 are also shown in Figure 5.7. The formation of the epoxide could well be the key to both acute and chronic toxicity, and those animals which fail to produce it are relatively resistant to both. Those animals which produce the epoxide, but do not effectively metabolize it further, may be at the highest risk to the carcinogenic activity of aflatoxin B_1, but relatively resistant to the acute toxicity. Those animals which not only produce the epoxide but also effectively remove it with a hydrolase enzyme, thus producing a very reactive hydroxyacetal, are most sensitive to the acute toxicity. The epoxide of aflatoxin B_1 is known to react with guanine residues in DNA, and can cause subsequent depurination, while the hydroxyacetal derivative reacts with proteins through such residues as lysine. The parent molecule may thus be seen as a very effective delivery system, having the right properties for absorption from the gut and transmission to the liver and other organs of the body. It is, however, the manner in which it is subsequently metabolized *in vivo* which determines the precise nature of the animal's response. Such information as is available about the metabolic activity in the human liver confirms that people of different races are going to be intermediate in sensitivity to the acute toxicity and may show some sensitivity to the chronic toxicity of aflatoxin B_1, including carcinogenicity.

Another aspect of the possible significance of aflatoxins in human health has been exposed by studies carried out by Professor Hendrickse and his colleagues at the University of Liverpool of the levels of aflatoxins, and the metabolite

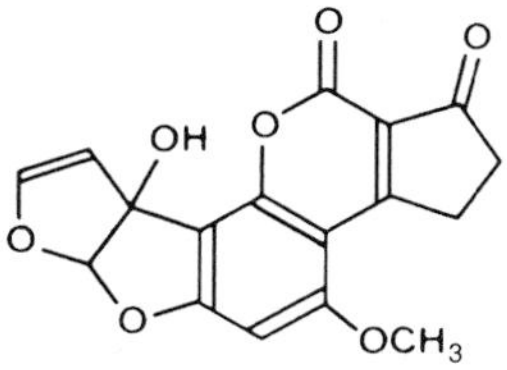

Aflatoxin M_1

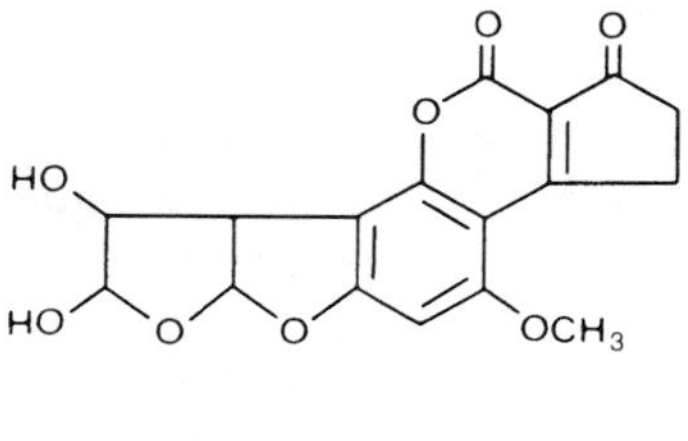

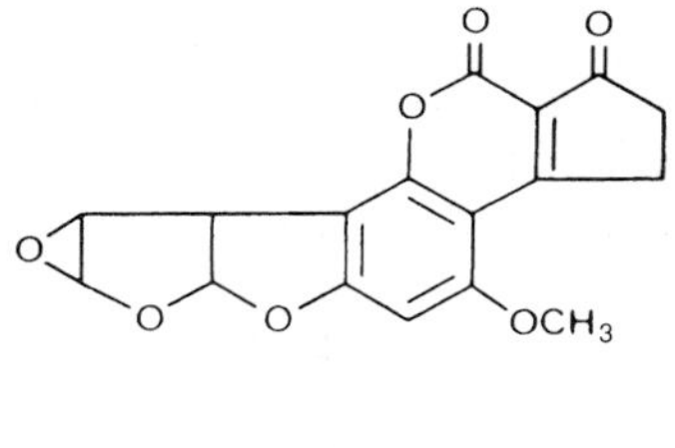

2 - 3 Dihydroxyaflatoxin

2 - 3 Epoxyaflatoxin

Figure 5.7 The structure of aflatoxin M_1 and other products of the metabolism of aflatoxin B_1.

aflatoxicol, in the blood and other body fluids of children in the tropics. It has been proposed that aflatoxin in the diet of children is associated with the occurrence of kwashiorkor. Indeed, a number of studies have demonstrated that very young children may be exposed to aflatoxins even before they are weaned, because mothers consuming aflatoxin in their food may secrete aflatoxin M_1 in their milk. There is no doubt about the potential danger of aflatoxin in food, and every effort should be made to reduce or eliminate contamination. However, it may be very difficult to totally eliminate aflatoxin from the diet of most people of the world and, as a first stage in making a risk assessment, an estimate of the average daily intake for an adult American has been made. For aflatoxin B_1 it has been estimated to be of the order of 20 ng/kg body weight (mainly from maize and maize products) and for aflatoxin M_1 of the order of 0.8 ng/kg body weight (from milk and milk products). It should be noted that aflatoxin M_1 is less carcinogenic than B_1 by a factor of at least 100.

Although correct post-harvest storage conditions and prevention of mould growth and mycotoxin biosynthesis should be the highest priority, it has been necessary to accept that mycotoxin contamination may be inevitable in some commodities under some conditions. In this case a reasonable question is: can food be decontaminated? There is no general answer, several mycotoxins being remarkably stable in the complex food matrix even when subjected to cooking. Only in the case of aflatoxin have large-scale decontamination methods been developed; the most favoured process is treatment with ammonia gas under relatively mild conditions of temperature and pressure after increasing the moisture content of the commodity. At present such decontaminated commodities can only be used for incorporation into animal feeds.

5.4.2 Other *Aspergillus* toxins

The genus *Aspergillus* is very rich in the production of toxic metabolites, but in this brief review it is appropriate to mention only two more. Sterigmatocystin, a precursor in the biosynthesis of aflatoxins, is produced by a relatively large number of moulds, but especially by *Aspergillus versicolor*. It is not considered to be as acutely toxic or as carcinogenic as aflatoxin, but it is likely to be quite widespread in the environment and has been isolated from a number of human foods. Cyclopiazonic acid gets its name because it was first isolated from a mould which used to be called *Penicillium cyclopium* (now known as *P. aurantiogriseum*), but is has subsequently been isolated from *A. versicolor* and *A. flavus*. In the latter it is formed primarily in sclerotia, and there has always been a suspicion that some of the symptoms ascribed to the ingestion of food contaminated by *A. flavus* may be due to the presence of this compound, as well as aflatoxins.

In parts of India a disease known as kodua poisoning occurs following the consumption of kodo millet (*Paspalum scrobiculatum*), which is both a staple food and an animal feed. *A. flavus* and *Aspergillus tamarii* have been isolated from incriminated samples of millet, and both species are able to synthesize cyclopiazonic acid. Poisoning by cyclopiazonic acid of cattle and humans is associated with symptoms of nervousness, lack of muscle coordination, staggering gait, depression and spasms and, in humans, sleepiness, tremors and giddiness which may last for 1–3 days. Some of these symptoms are reminiscent of a range of disorders in intensively reared farm animals, collectively referred to as 'staggers', in which complex indole alkaloid metabolites known as the tremorgens are implicated. One of these metabolites, aflatrem, is also produced by some strains of *A. flavus*.

5.5 OTHER FUNGI

Ergotism has been documented as a human disease since the Middle Ages, but its aetiology remained a mystery until the mid-19th century, when it was demonstrated that it was caused by a fungus, *Claviceps purpurea*. This fungus is, in fact, a very specialized parasite of some grasses (including cereals) and, as part of its life cycle, the tissues of infected grains become completely replaced by fungal mycelium, producing a tough, purple-brown structure referred to by mycologists as a sclerotium, but also known as an ergot. The biological function of such structures is to survive adverse conditions; in this case it is designed, like the seed of the host plant, to over-winter in order to germinate in the following spring. These ergots contain alkaloid metabolites such as ergotamine which are incorporated into the flour and eventually into bread made from the harvested contaminated cereals.

Today, the toxicity of the ergot alkaloids is well understood and ergotism, or St Anthony's fire, is infrequent in human beings. One aspect of their activity is to cause a constriction of the peripheral blood vessels, leading in extreme cases, to fingers and toes becoming gangrenous and necrotic. Different members of this family of mould metabolites may also have profound effects on the central nervous system and they may also stimulate smooth muscle activity. Indeed, when purified and used at controlled doses, the ergot alkaloids form a valuable collection of medicines. For example, ergometrine is used for the control of post-partum haemorrhage.

Plant–fungal interactions may be very complex, and there are instances where a toxic plant metabolite is produced in response to fungal attack. Such a situation occurs when the sweet potato, *Ipomoea*, is damaged by certain plant pathogens. It responds by producing the phytoalexin ipomeamarone, an anti-fungal agent produced to limit fungal attack, which is also a hepatotoxin to mammals. Further complexity arises when other moulds, such as *F. solani*, degrade ipomeamarone to smaller molecules such as ipomenol, which can cause pulmonary oedema.

5.6 CONCLUSIONS

It is important to recognize that fungal metabolites which may cause health problems as a result of ingestion in food may enter the body by other routes and, in the case of the study of mycotoxins, it is an over-simplification of their potential importance to think of them only as food poisoning agents. Thus, during the past few decades there has been an increasing concern about the nature of the health hazards associated with living in damp housing, and a number of laboratories in several parts of the world have been studying the possibility that volatile mycotoxins, or mycotoxins carried in air-borne fungal spores, may be implicated.

Three of the most important mycotoxins, aflatoxin, ochratoxin and T-2 toxin, are immunosuppressive, but quite distinct in their activity against the immune system. Aflatoxin, which can influence protein biosynthesis indirectly by inhibiting transcription, suppresses cell-mediated responses, impairs complement (C_4) activity and suppresses IgG and IgA, but not IgM. Ochratoxin, on the other hand, influences protein biosynthesis by inhibiting phenylalanyl tRNA synthetase, has no effect on complement and suppresses IgG and IgM, but not IgA, and inhibits macrophage migration. T-2 toxin has a more direct effect on protein biosynthesis by inhibiting translation through binding with a specific site on the eukaryote ribosome. It brings about a reduction of complement (C_3), suppresses IgA and IgM, but not IgG.

One consequence of these distinct modes of activity is that mixtures of such mycotoxins are likely to be synergistic in activity, and this has been shown experimentally in the case of aflatoxin and T-2 toxin. This observation is significant in that a food which has gone mouldy will probably be infected by several species of mould, and may thus be contaminated by several distinct mycotoxins belonging to different groups. It is generally the case that the mycotoxins are produced in families of related compounds (there are several aflatoxins and ochratoxins as well as a large number of trichothecenes), but synergism is most likely to occur between pairs of mycotoxins from different families.

Although there can be no doubt about the potential for mycotoxins in food to cause illness and even death in humans, there are far more overt mycotoxicoses in farm animals throughout the world which have a major impact on the economy through losses in productivity.

Recognition of the potential to cause harm to the human population by the imposition of maximum tolerated levels of mycotoxins such as aflatoxin, can also have a major impact on economics by rendering a commodity unacceptable in national or international trade. Thus, a major problem occurred for Turkey, the world's most important exporter of dried figs, during the Christmas of 1988. Several European countries imposed a ban on the import and sales of dried figs following the demonstration of aflatoxin in 30% of samples of figs analysed. Fearing the loss of 50 000 jobs in the fig-drying and packing industry, Turkey was vigorous in her diplomatic efforts to have the bans lifted. This was done fairly soon after they had been imposed and an international symposium on 'Dried figs and aflatoxins' was held in Izmir, Turkey, during April of 1989. In 1980, nearly 66% of random samples of maize from North Carolina had concentrations of aflatoxins in excess of 20 µg/kg, giving rise to an estimated loss to producers and handlers of nearly 31 million dollars. It is rare that the losses and costs arising from mycotoxin contamination can be calculated, but these two isolated and very different examples indicate that on a worldwide basis they must be considerable. In both these examples aflatoxin was probably formed in the commodity during growth and development in the field. Under these conditions aflatoxin formation is usually relatively low, and in neither case was there any evidence of harm to human beings.

Three reports on mycotoxins in the UK have been prepared by the Steering Group on Food Surveillance for the Ministry of Agriculture, Fisheries and Food (Food Surveillance Papers Nos 4, 18 and 36). They describe the occurrence of low levels of mycotoxins in foods, but conclude that there is negligible mycotoxin contamination in the diet of most people in the UK, with the exception of aflatoxins in samples of nuts and nut products. One result of the first report was action which considerably reduced the levels of aflatoxin M_1 in milk and the third report documents the background to setting legislative levels for aflatoxin B_1 in figs, nuts and their products as well as the advisory level for patulin in apple juice in the UK. Table 5.4 indicates the range of commodities which might cause some concern in Europe. However, it is when commodities are improperly stored in rural communities in developing countries that really high concentrations of mycotoxins may be formed, and it is in these situations that human suffering can occur.

Table 5.4 Foodstuffs associated with mycotoxins in Europe

Mycotoxin	Edible nuts	Cereals	Dried fruits	Fruit juices	Meat products	Milk
Aflatoxin B_1	+	+	+	−	−	−
Aflatoxin M_1	−	−	−	−	−	+
Deoxynivalenol	−	+	−	−	−	−
Ochratoxin	−	+	−	−	+	−
Patulin	−	−	−	+	−	−
Zearalenone	−	+	−	−	−	−

6 Viruses and protozoa

A.R. Eley

In up to 10% of all reported general community outbreaks of food poisoning in the UK, bacteria cannot be identified as the causative organisms. In many of these instances it is becoming increasingly evident that viruses are playing an important role; hospital outbreaks of viral gastroenteritis are now known to be relatively common. One of the reasons for the lack of detailed knowledge about viruses that have been implicated in food poisoning is that traditionally they have only been considered after bacterial causes have been excluded; this often means that there is no food left for examination. Another problem has been the difficulty of observing or isolating viruses from foods: they may vary greatly in composition, with each requiring a different extraction procedure. Furthermore, techniques for isolating viruses are more complex and more time-consuming than those used for isolating bacteria. Unfortunately, electron microscopy has only a limited level of sensitivity (approximately 10^5–10^6 viral particles/g of food), whereas with certain viruses we know that the infective dose may be as low as 10 to 100 particles/g. It is hoped that new methods for detecting viruses, including DNA probe and amplification technology, may have much greater sensitivity and provide new information on the natural history of food-borne viral gastroenteritis.

Although the role of viruses in gastroenteritis is under investigation, so far only a relatively small number of different organisms have been frequently associated with food poisoning. These are discussed below in order of importance:

1. small round structured virus (SRSV) group;
2. rotavirus;
3. astrovirus;
4. other viruses.

Although protozoa are not considered to be classical agents of food poisoning, recent evidence has indicated a food mode of transmission for some cases of cryptosporidiosis, as well as a small number of outbreaks of giardiasis. Amoebic dysentery caused by *Entamoeba histolytica* is usually a water-borne infection, but it has potential for food-borne transmission, especially in endemic areas of the world such as the Indian sub-continent and tropical Africa.

Improvements in methodology and increased awareness should allow us to look more closely for protozoa, so that we may appreciate their true role in the causation of gastroenteritis.

Like viruses, these protozoa cannot replicate in food, but their cysts may remain infectious in foods for prolonged periods, constituting a potential hazard. Furthermore, only a small number of cysts are required to cause disease (i.e. they have a low infective dose).

6.1 SRSV GROUP

Unfortunately, the taxonomy and nomenclature of this group of viruses is confusing, misleading and subject to rapid change, and differs between the UK and USA.

Perhaps the best known virus in this group is the Norwalk agent, which is an important cause of gastroenteritis in older children and adults, usually during the winter months ('winter vomiting disease'). This virus was initially described in 1972 and derived its name from an outbreak of acute gastro-enteritis in a secondary school in Norwalk, Ohio. In the UK, however, this virus is known as SRSV. In addition to the Norwalk virus, there are a small number of related viruses, which may be described as structurally similar but antigen-ically different. These are called Norwalk-like viruses (NLVs), or other SRSVs, and these include Hawaii virus, Snow Mountain virus, Montgomery County virus and Taunton virus; the names usually relate to the location of the initial outbreak.

At present the nucleic acid type, which is an important characteristic in viral classification, is not known for all SRSVs. However, recent reports suggest that the Norwalk and Snow Mountain viruses contain a single major structural protein similar to that found in human caliciviruses. This finding has prompted the suggestion that the Norwalk and related viruses may be possible candidates for consideration as members of the Caliciviridae. Briefly, the major features of the Caliciviridae are: non-enveloped, roughly spherical and approximately 35–39 nm in diameter, capsid has a single major poly-peptide (60 000–71 000 Da molecular weight), one molecule of single-stranded RNA. These viruses are not sensitive to lipid solvents or mild detergents, but are inactivated at pH 3–5.

Viruses within the SRSV group are generally described as virions with a diameter of 27–32 nm (Figure 6.1). This makes them slightly smaller than human caliciviruses and slightly larger than parvoviruses (18–26 nm). The latter distinction should be noted, as SRSVs have been and still are called 'parvovirus-like particles'. This name may now be even more confusing, as we know that parvoviruses contain single-stranded DNA, in contrast to the RNA content of caliciviruses.

One of the major problems with SRSVs has been the difficulty of *in vitro*

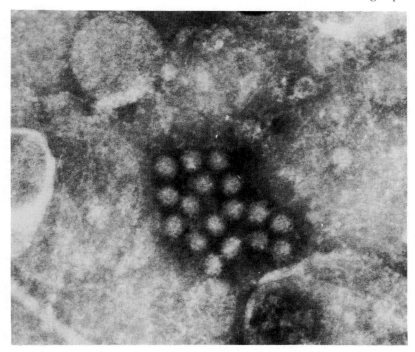

Figure 6.1 SRSV in faecal extract (× 200 000).

cultivation, which means that few of the viruses have been compared or characterized in detail; this has resulted in slow taxonomic progress.

6.1.1 Pathogenesis

At present disease mechanisms are poorly understood, although villous blunting and inflammation do occur: it is known that only a small infective dose is required (10–100 virions/g of food). It has also been established that these viruses are resistant to gastric acid, unlike some human caliciviruses, which would be inactivated in the stomach (see above).

6.1.2 Clinical features and prognosis

Symptoms of SRSV infection are typically nausea, vomiting, abdominal pain, malaise and a low-grade fever. Diarrhoea may also be present, though often in a mild form. The disease has an incubation period of between 24 and 72 hours (Table 6.1).

Table 6.1   Clinical features of the illnesses produced by the major viral and protozoal causes of food poisoning

	SRSV	Rotavirus	Crypt. parvum	G. lamblia
Incubation time (h)	24–72	24–168	4–14 days	1–3 weeks
Duration of				
illness (h)	24–48	48–72	2–14 days	1–4 weeks
Vomiting	+	+	+	−
Nausea	+	−	+	+
Diarrhoea	±	+	+	+
Abdominal pain	+	−	+	+
Fever	+	±	±	−

−, rarely seen; ±, sometimes seen; +, often seen.

The disease is usually self-limiting, of 24–48 hours' duration, with no long-term sequelae.

6.1.3   Incidence and epidemiology

Due to the limitations on the detection and identification of these viruses, it is impossible to quote figures on the incidence of this type of food poisoning. However, it is generally believed that this group of viruses may be the most common non-bacterial cause in the UK and the USA. It is also thought to play a significant role in Australia and Japan. The epidemiology of infection is further complicated by the fact that some patients may be asymptomatic excreters for a period of time. Symptomless food-handlers may be an important source of food contamination.

Epidemics often occur in families, communities, schools, camps, institutions and cruise ships and spread rapidly, as these viruses have a low infective dose.

6.1.4   Ecology and foodstuffs

SRSV food poisoning has been associated with certain foods, especially shellfish. This is because aquatic bivalve molluscs (clams, mussels, oysters) filter and concentrate microscopic organisms from faeces-contaminated water into their own bodies. When eaten raw or improperly cooked, these infected shellfish can then cause disease. However, these viruses are not capable of multiplying in such foods.

6.1.5   Control

Since our knowledge of these organisms is limited, the only practical means of control are adequate cooking of foods, the prevention of cross-contamination and other good food hygiene practices.

SRV group

Another group of viruses with a similar terminology are the small round (feature-less) viruses (SRVs). These were first described in 1978 and are generally smaller in size (20–30 nm in diameter) than SRSVs. As with the SRSVs, their nucleic acid type is unknown, they have a low infective dose, and disease is often associated with consumption of raw or improperly cooked clams or oysters. However, their true significance in causing disease is still not fully established.

6.2 ROTAVIRUSES

This group of viruses was first discovered in 1973; the first types to be isolated are now known as group A rotaviruses. These are the most common cause of gastroenteritis in infants worldwide, and in developing countries they account for high levels of infant mortality. Usually, rotaviruses are responsible for a winter disease, with a peak in autumn and early winter.

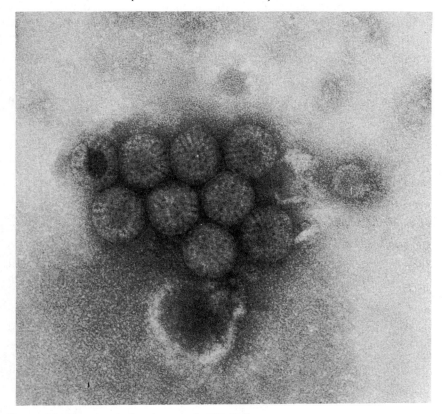

Figure 6.2   Rotavirus in faecal extract (× 200 000).

Rotaviruses are described as reovirus-like particles of approximately 70 nm in diameter, which have a wheel-like shape, a double-shelled capsid structure, and contain double-stranded RNA (Figure 6.2).

There is still a certain amount of controversy about whether rotavirus infections can be food-borne, which has not been helped by the fact that these viruses grow poorly or not at all in cell-culture systems. Food is almost certainly not the principal mode of transmission, which is more likely to be faecal-oral, a suggestion supported by the fact that rotaviruses can survive on human hands. However, there is some evidence that food contamination may be responsible for single isolates and a number of outbreaks.

6.2.1 Pathogenesis

Little is known about their pathogenesis apart from the fact that these viruses can infect the mucosa of the jejunum and ileum, causing desquamation, villous atrophy and delaying enterocyte differentiation. This could then lead to carbohydrate malabsorption and finally diarrhoea. It should be noted that only a very low infective dose is necessary to cause disease, while the virus can be excreted in large numbers in faeces, up to 10^{10} virions/g. Maximum viral shedding in stools tends to occur from 2–5 days after the onset of diarrhoea.

We do not fully understand yet why many adults with infection are asymptomatic, or why others may display a wide variety of symptoms. A number of factors may be involved, including a pre-existing immunity, differences in immune response, infections with other pathogens, differences in inoculum size, and host factors such as stress or drug ingestion. A further complication is that these viruses may often be found in large numbers in the stools of healthy neonates, but are then associated with symptomatic disease in children older than 6 months.

6.2.2 Clinical features and prognosis

This type of food poisoning is characterized by vomiting (sometimes projectile) and diarrhoea; there may be an associated upper respiratory tract infection. If fever is also present then this may indicate a more severe illness. The incubation period may vary from 1–7 days, but is usually between 2–4 days.

A complete recovery usually occurs within 2–3 days, although fatalities may very occasionally be reported. Deaths tend to occur among young children and geriatric patients; both groups tend to suffer more severe symptoms, including dehydration.

6.2.3 Incidence and epidemiology

Group A rotaviruses, so-called because of the presence of a group A specific antigen, are the most common isolates from Western Europe and the USA.

Recently, other serologically distinct groups have been discovered, which have a rather different geographical distribution. The group B viruses have been responsible primarily for larger-scale epidemics among infants and adults in China. In 1989 there was a report from Japan of the first large outbreak due to group C rotaviruses: out of 3102 people, 675 became ill. Previously, group C viruses had only been implicated in a few sporadic cases. Furthermore, in the last few years there have been reports of viruses found in patients with gastroenteritis which have rotavirus morphology but lack the common group specific antigen; these have been called atypical rotaviruses or pararotaviruses.

In institutional outbreaks in hospitals, nursing homes, day-care centres and schools, epidemics are characterized by a large-scale sudden onset of presenting symptoms (see above), usually of short duration.

For epidemiological purposes, rotaviruses have been divided into groups based on major distinct antigen specificities. Subgroups have been based on the major inner capsid protein VP6 which has a molecular weight of 45 000 Da, and serogroups divided on the basis of the major outer capsid glycoprotein VP7 which has a molecular weight of 35 000 Da. In addition, analysis by polyacrylamide gel electrophoresis, which allows separation in an electrical field according to molecular size, has shown that the genome consists of 11 segments of double-stranded RNA which vary from strain to strain, resulting in a characteristic pattern or electropherotype.

6.2.4 Ecology and foodstuffs

Infection is thought to occur by ingestion of faeces-contaminated food (such as seafood) or water, although no specific foodstuffs are known to be particularly hazardous at present. What is known is that rotaviruses are fairly stable in groundwater and sewage, and are resistant to many commonly used chemical disinfectants.

6.2.5 Control

For general measures see section 2.6. It must also be stressed that because this disease can be transmitted in a number of different ways and that the infective dose is low, strict isolation and barrier nursing are recommended to prevent hospital outbreaks. Moreover, because this virus is relatively resistant to disinfectants it will spread readily when hygiene is inadequate.

6.3 ASTROVIRUSES AND OTHER VIRUSES

These viruses were first described in 1978. They are 28–30 nm in diameter and have a characteristic five- or six-pointed, star-shaped surface pattern. Their nucleic acid type is single-stranded RNA. At least five serotypes have been recognized in the UK.

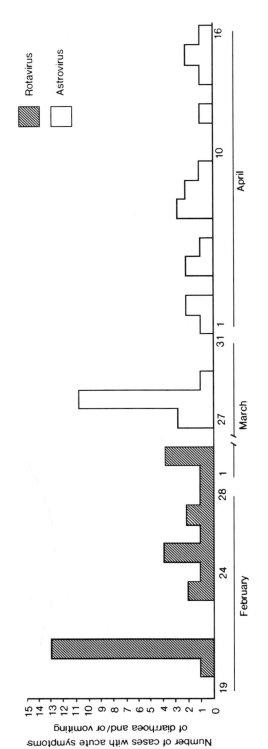

Figure 6.3 Cases of gastroenteritis on a geriatric ward, caused by rotavirus followed by astrovirus. After Lewis, D.C. *et al.* (1989) *Journal of Hospital Infection*, **14**, 9–14.

Table 6.2 Other viruses associated with gastroenteritis

Virus	Year described	Virion diameter (nm)	Nucleic acid type*
Enteric coronavirus	1975	100–150	ss RNA
Human calicivirus	1976	27–32	ss RNA
Enteric adenovirus	1979	70–80	ds DNA

*ss, single-stranded; ds, double-stranded.

It is generally believed that these viruses are less pathogenic in adults than those belonging to the SRSV group. In two recorded outbreaks it was interesting to note that astrovirus infection occurred a few weeks after an outbreak of a different viral gastroenteritis. In one instance, of those patients previously infected by rotavirus, 50% were infected by astrovirus in the second outbreak. This suggests that infection with one gastroenteritis virus may predispose to infection by another (Figure 6.3).

Astroviruses have been detected in normal and diarrhoeal faeces from animals and humans, although they are difficult to cultivate *in vitro*. Partial characterization, however, has been made possible by the extremely large numbers of virions shed in the faeces.

Few outbreaks of gastroenteritis in adults involving astrovirus infection have been reported, although they are more common among school children and in paediatric wards. In one recent outbreak it was suggested that food contamination provided a plausible explanation for the pattern of infection, although no proof was found of virions in the food.

A number of other viruses have been associated with gastroenteritis (Table 6.2), and at least one, enteric adenovirus, may be excreted in large numbers in faeces. However, these other viruses have been difficult to cultivate *in vitro* and as yet no food-associated outbreaks have been reported. Further studies and newer technological advances should allow us to understand them more fully.

6.4 *CRYPTOSPORIDIUM PARVUM*

Cryptosporidium is a genus of protozoa which is pathogenic for humans and other animals.

6.4.1 Pathogenesis

Infection follows the ingestion of a small number (probably <10) of oocysts (cysts forming sporozoites), typically 4–6 µm in size. These banana-shaped motile sporozoites are released in the small intestine, where they adhere to enterocytes of the villi and develop into trophozoites beneath the cell membrane. Fertilization of macrogametes may follow, which results in the production of oocysts. Two types of oocyst can be formed: thin-walled oocysts which release sporozoites into the host's intestine, causing re-infection ('auto-infection') of the host, and acid-fast, thick-walled oocysts which constitute approximately 80% of the total, and are released in the faeces (Figure 6.4).

The precise mechanism of pathogenesis is unknown, although the diarrhoea produced is of a secretory nature (with possible involvement of an enterotoxin), with damage to the villi and some resulting malabsorption. However, invasion beyond the host cell membrane does not usually occur.

6.4.2 Clinical features

Symptoms of cryptosporidiosis in patients without immunodeficiency are diarrhoea of 2–14 days' duration, sometimes accompanied by an 'influenza-like' illness and fever. Additional features are nausea, vomiting and abdominal pain, usually after an incubation period of 4–14 days.

In immunocompromised patients, especially those with AIDS, much more

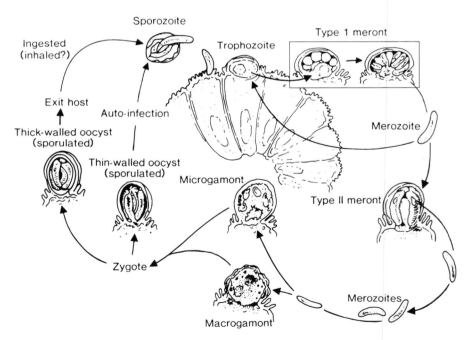

Figure 6.4 Life cycle of *Cryptosporidium parvum*. (Source: Casemore, D.P. (1990) *Lancet*, **336**, 1427–32.)

severe symptoms of diarrhoea, malabsorption and weight loss have been seen. In Africa the condition known as 'slim disease' refers to AIDS patients who suffer the devastating consequences of cryptosporidiosis.

6.4.3 Prognosis

In children, where the symptoms are less severe, the illness is usually self-limiting. However, in immunocompromised patients the illness may be much more severe, with some fatalities.

Rehydration is important in severe cases such as malnourished children and AIDS patients. Chemotherapy is controversial, however, as there have been no definitive studies of proven efficacy, and because of this, cryptosporidiosis is extremely difficult to treat. Certain antibiotics have been shown to shorten the course of infection although relapses may occur. Furthermore, anti-peristaltic agents are usually successful in easing the symptoms of diarrhoea.

6.4.4 Incidence and epidemiology

The first human case of cryptosporidiosis caused by *Crypt. parvum* was described in 1976. This organism is now recognized as an important cause of diarrhoeal disease (including traveller's diarrhoea) worldwide. Initially, infections were associated with immunocompromised patients, especially those with AIDS, although nowadays most infections are seen in young children; infections are generally less prevalent in older children and adults.

Worldwide, there appears to be a much greater incidence of cryptosporidiosis in children in third-world countries than in Western Europe. Since these children are more likely to be malnourished, the disease is of greater severity and can lead to life-threatening dehydration. At the same time there appears to be a seasonal distribution of infections, with a predominance in late spring/early summer.

As food-borne transmission of cryptosporidiosis has only recently been suggested, there are no reported figures for the incidence of this type of food poisoning. However, in 1993 the largest water-borne disease outbreak (approximately 400 000 people) ever recognized in the USA was caused by *Crypt. parvum*.

6.4.5 Ecology and foodstuffs

Until recently, cryptosporidiosis was thought to be a zoonotic disease, with cattle and other farm animals being implicated, but is now well established that the disease may also be transmitted from person to person. Food transmission is thought to be linked with untreated milk and processed meats, as well as contaminated drinking water as a food component.

6.4.6 Control

Fortunately, pasteurization of milk renders the oocysts non-infective. In addition, oocysts are destroyed by heating and freezing, although they are resistant to many disinfectants, including chlorine.

6.5 *GIARDIA LAMBLIA*

Giardia lamblia is a well-recognized cause of diarrhoeal illness, and is the most common gastroenteric parasite of humans in the Western World. This flagellate protozoan (synonyms: *G. intestinalis; Lamblia intestinalis*) has both trophozoite and cystic stages and usually occurs in immunocompetent patients.

6.5.1 Pathogenesis

As with cryptosporidium, only a few cysts (<10) need to be ingested to cause infection. The cyst is oval (Figure 6.5), 7–14 μm in length and when mature has four nuclei and a diagonal axoneme (at this size the cyst is approximately 300 times larger than SRSVs).

Excystation and multiplication take place in the small intestine, particularly in the jejunum, where a four-nucleated trophozoite emerges and immediately divides by binary fission. Morphologically, the trophozoite is pear- or kite-shaped, measures 10–20 μm in length, has two oval nuclei and eight flagella, and has a prominant ventral sucking disc (Figure 6.5). The latter enables the

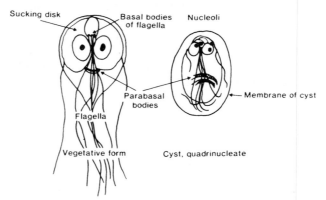

Figure 6.5   Vegetative and cyst forms of *Giardia lamblia.*

parasite to adhere to the intestinal wall of the host, where it may act as a mechanical obstruction of the intestinal mucosa. In some patients the trophozoites become attached to the villi and cause mucosal damage. In contrast to cryptosporidia, giardia not only attach to the epithelium but also penetrate beyond it.

Although the mechanism by which *G. lamblia* induces diarrhoea is not clear, it has been reported that it produces a cholera toxin-like protein (which cross-reacts with antibodies to cholera toxin) and activates adenylate cyclase in a manner identical to cholera toxin.

6.5.2 Clinical features and prognosis

Symptoms of giardiasis include diarrhoea, which may develop into steatorrhea (fatty stools) if heavy infestation results in malabsorption of fat in the intestine. Nausea and abdominal pain are often present, although vomiting and the presence of a fever are rare. The incubation period is usually 1–3 weeks after the ingestion of contaminated food.

Infection is usually self-limiting after a period of a few weeks. In severe symptomatic infestations antimicrobial treatment may be recommended. Metronidazole is still probably the drug of choice, although the combination of metronidazole and mepacrine has been suggested when metronidazole alone fails to eradicate the infection. A single dose of tinidazole may be used as an alternative.

6.5.3 Incidence and epidemiology

At least three food-borne outbreaks have been described, all of which have occurred in the USA. Home-canned salmon was implicated in the first outbreak, with noodle salad possibly implicated in the second. In the third outbreak no single food item was thought to be responsible, although there was an association between illness and eating sandwiches and uncooked food.

It is possible that similar outbreaks have occurred elsewhere and have gone undetected because of a number of factors, including problems in establishing a laboratory diagnosis, the long incubation period and a tendency not to consider giardia as a possible agent of food poisoning. Giardiasis is usually acquired by drinking inadequately treated contaminated water or by person-to-person spread by the faecal–oral route, and is transmitted by the ingestion of cysts.

In contrast to cryptosporidiosis, malnutrition does not predispose to a greater severity of infection. However, there is an increased prevalence of giardiasis in some immunocompromised patients, especially those with hypogamma-globulinaemia, where symptoms may be more severe or prolonged.

6.5.4 Ecology and foodstuffs

There appears to be no association between infection and consumption of particular foodstuffs, although very few outbreaks have been reported (see above). It is more likely that problems will arise whenever foods such as uncooked vegetables come into contact with water contaminated with faecal material, which is usually the source of infection.

6.5.5 Control

Normal cooking procedures will kill any infective oocysts present. However, raw food and cross-contamination are hazards, especially when there has been contact with faeces-contaminated water. Obviously, good personal hygiene practices will also prevent direct faecal contamination in foods.

Of particular importance is the fact that the cyst stage is resistant to the chlorine concentrations used in most water treatment plants. Therefore, adequate water treatment should include filtration combined with chemical treatment.

7 Laboratory diagnosis

A.R. Eley

In this chapter we will discuss how a laboratory diagnosis of food poisoning can be achieved using conventional methodology and by newer methods.

As this is not a laboratory manual, we will not be describing detailed methods for the detection and identification of all food poisoning micro-organisms from both clinical specimens and food samples. Instead, we will concentrate on the principles behind the procedures, giving examples wherever possible; only the isolation and presumptive identification of the ten most important bacteria which cause food poisoning will be discussed, along with viruses, mycotoxic fungi and protozoa. In a similar vein, although we realize that one of the final objectives of laboratory diagnosis is the definitive identification of an organism of interest, it is not our intention to provide a comprehensive guide to the identification of these bacteria: this information can be found in the appropriate manuals. Our aim is rather to remind the reader of distinctive characteristics which may be suggestive of a particular bacterium, before a final, comprehensive procedure is performed.

Finally, we will outline the epidemiological typing methods which can help in tracing the spread of disease and the sources and reservoirs of infection.

7.1 CLINICAL SPECIMENS AND FOOD SAMPLES

Firstly, we need to identify the causative organism from clinical specimens, so that the patient may receive the appropriate therapy as quickly as possible. Secondly, it is important to trace the food source of the disease, so that measures can be taken to halt the possible spread of infection, and so that epidemiological investigations can be initiated at an early stage.

7.1.1 Clinical specimens

A preliminary clinical diagnosis may be made on the basis of the presenting clinical features, the incubation period and the type of food consumed; one or more of these factors may be extremely characteristic of a particular causative organism. However, for a definitive laboratory diagnosis, clinical specimens will need to be collected. As with other microbial diseases, there should be careful collection of the appropriate specimens, such as stools or vomit, during the acute phase of the illness. It is also essential that the specimens be transported to the laboratory without delay, so that laboratory investigations can be started at the earliest possible opportunity.

In the majority of food poisoning cases, only stools are examined for the most important micro-organisms. Enrichment procedures are followed by culture and plating of the bacteria on to a solid medium, which may or may not have differential and/or selective properties (e.g. MacConkey's agar). Occasionally, vomit may be tested for the enterotoxin of *Staph. aureus*, cultured for *B. cereus*, and examined under the electron microscope for rotaviruses (Table 7.1). For those bacteria that may enter the bloodstream and produce a systemic disease (e.g. salmonella), it would be advisable to take blood samples for culture. In addition, in botulism cases a clotted blood sample for serum separation can be taken to detect botulinum toxin.

If protozoa are suspected, a suitable concentration method followed by light microscopy is useful. The greater magnification achieved by electron microscopy is used for the detection of viruses, many of which, e.g. rotaviruses, can be recognized by their characteristic morphological appearance. These viruses in particular may also be detected using serological techniques specific for their antigens. Finally, proof that a particular bacterium is the causative agent of a

Table 7.1 Testing of clinical specimens and food for food poisoning organisms

Specimen analysis	Organism
Vomit	
Culture	*B. cereus Cl. botulinum*
Toxins	*Staph. aureus, Cl. botulinum*
Electron microscopy	Rotaviruses
Stools	
Culture	*B. cereus, Cl. perfringens,* *Cl. botulinum, Campylobacter* spp., *E. coli, Salmonella* spp., *Vibrio* spp., *Yersinia* spp.
Toxins	*Cl. perfringens, Cl. botulinum,* EHEC, ETEC
Electron microscopy and/or ELISA	Rotaviruses
Light microscopy	*Giardia, Cryptosporidium*
Blood culture	*Salmonella* spp., *Campylobacter* spp.
Serum	
Toxin	*Cl. botulinum*
Foods	
Culture	All food poisoning bacteria
Toxins	*Staph. aureus, B. cereus* (diarrhoeal toxin), *Cl. botulinum*

disease can be established by identifying its principal mode of pathogenesis, i.e. by toxin detection.

7.1.2 Food samples

The major concern of food microbiologists, as of their clinical counterparts, is the rapid identification of pathogens. In order to facilitate this process, food samples should be analysed as soon as possible, or else stored under conditions which will arrest active microbial growth. Frozen foods should preferably be thawed in the refrigerator, and dry foods held at ambient temperature pending analysis.

Although some methods used in clinical microbiology can be applied to the microbiological analysis of foods, there are certain differences between clinical specimens and food samples, which often require quite separate procedures. One important difference is the heterogeneous and variable nature of foods, in terms of composition and microflora. There are many more classes of food than the small number of types of clinical specimens usually sent for examination. In general, there is probably also a greater diversity of microbial species in foods than in clinical specimens. Furthermore, those bacteria responsible for disease causation may be found in large numbers in clinical specimens, but will probably only be present in very low numbers in foods. Low numbers of organisms may, however, be extremely significant in food poisoning: only low infective doses of salmonella and campylobacter may be required to cause disease. This underlines the need for proper selective enrichment procedures to reduce the level of false-negative results.

The recovery of pathogenic strains of bacteria from food may be seriously impaired by physiologically demanding cultural conditions, such as elevated or lowered temperatures, and by the addition to the culture medium of selective agents such as detergents, toxic inhibitors, or antibiotics. In some cases a resuscitation step may be needed, as certain bacteria may be injured by the treatment of food during preparation (e.g. heating or freezing), or by exposure to components of the food product such as preservatives. It is therefore essential to give careful consideration to the design of schemes for recovering micro-organisms from food, taking into account the nature both of the organisms which are potentially present and of the foodstuff(s) concerned.

7.2 CONVENTIONAL METHODOLOGY

The conventional methods of laboratory diagnosis usually consist of utilizing a range of culture media to encourage growth of the pathogen, even in the presence of many other contaminating bacteria. When the pathogenic bacteria have been isolated in pure culture, it is then possible to perform simple tests so that a presumptive, and then final identification can be made. If outbreaks need investigating, further traditional microbiological tests are usually necessary to type bacteria and to determine if they all belong to the same strain.

Alternatively, laboratories may look for toxins and mycotoxins, of bacterial and fungal origin respectively. Viruses are usually determined directly from clinical material either by antigen detection or by electron microscopy. Finally, light microscopy is often used in the examination of faecal material for protozoa and/or their cysts.

7.2.1 Culture media

Many different varieties of media are often recommended for particular situations and conditions. In what follows we give examples of appropriate media, with an indication of their composition, rather than a catalogue of precise details. It should be emphasized that on the whole the enrichment and solid media quoted as examples are designed primarily for clinical specimens; however, they may also be of use for growing the same bacteria from foods.

(a) Pre-enrichment media

The principal role of pre-enrichment (or non-selective enrichment) of foods is to provide damaged organisms with an opportunity to repair physiological lesions. Sublethal cell damage may have resulted from thermal processing of food, freezing, thawing, osmotic shock or prolonged storage of food. Typically, a pre-enrichment medium may be a buffered peptone water or nutrient broth. These are usually advantageous in the isolation of small numbers of salmonellae and listeria. It has been shown that the incubation period for pre-enrichment is critical: it must satisfy the minimum requirements for resuscitation of stressed or injured cells. With salmonellae, for example, the optimum time is thought to be 18–24 hours.

(b) Enrichment media

Enrichment procedures are relevant to stools as well as foodstuffs, and provide an important step in repressing large populations of competitive flora while encouraging growth of pathogens. Selectivity of the enrichment process may arise from inhibitory agents in enrichment media, pH or temperature of incubation, or a combination of these factors. Selective agents include antibiotics, and other inhibitory substances such as sodium biselenite (which inhibits most enteric bacteria except salmonellae), bile salts (most non-enteric bacteria are inhibited by bile salts), and sodium chloride, which in high concentrations is inhibitory to almost all pathogenic bacteria except staphylococci. With some organisms, e.g. salmonella and escherichia, it may be beneficial to incubate at a higher than normal temperature (normal being 37°C). In contrast, with psychrotrophic bacteria such as listeria and yersinia, a cold enrichment temperature of 4°C is useful, as at this temperature most other bacteria will not be able to multiply, although viability should be maintained. For certain bacteria such as *C. jejuni* (microaerophilic) and *Cl. botulinum* (anaerobic), which will not grow under, or are adversely affected by normal atmospheric conditions, it is essential to include a reducing agent in the enrichment medium to help diminish the damaging effects of oxygen.

(c) Selective media

These solid media are probably better described as selective-differential, as they often make it possible to recognize the characteristic colonial morphology of the desired bacterium when used for subculturing large mixed cultures. MacConkey's agar is an ideal example as it contains bile salts (see above), which are inhibitory, and lactose together with neutral red, which serve as substrate and indicator respectively. This allows easy differentiation of lactose fermenting and non-fermenting colonies, the former (e.g. *E. coli*) appearing red and the latter (e.g. salmonella) straw-coloured.

Other examples of selective media are shown in Tables 7.2 and 7.3. Like enrichment media, they often contain inhibitory substances such as sodium desoxycholate, lithium chloride and/or antimicrobials.

Table 7.2 Detection and identification of Gram-positive bacteria

| Bacteria | Culture media | | Presumptive identification |
	Enrichment	Solid	
B. cereus	Trypticase soy polymyxin broth (rarely needed)	5% horse blood agar	Mannitol-negative, lecithin hydrolysis
Cl. botulinum	Cooked meat broth	Egg-yolk agar	Lipase-positive, antitoxin precipitation
Cl. perfringens	Not needed	'Egg-yolk free' tryptose sulphite cycloserine agar (for spores)	Non-haemolytic, Nagler-positive
L. monocytogenes	PALCAM broth (polymyxin, acriflavin, lithium chloride, ceftazidime, aesculin, mannitol)	PALCAM agar	Growth at 4°C, tumbling motility at 22°C
Staph. aureus	Trypticase soy broth (+/–salt)	Baird–Parker agar (sodium lithium chloride and egg yolk-tellurite emulsion)	DNase positive, coagulase positive

Table 7.3 Detection and identification of Gram-negative bacteria

| Bacteria | Culture media | | Presumptive identification |
	Enrichment	Solid	
C. jejuni/coli	Brucella broth containing lysed blood, reducing agents and antibiotics	Selective medium containing vancomycin, polymyxin and trimethoprim	Micro-aerophilic, oxidase-positive
E. coli	Tryptose phosphate broth	MacConkey	Indole-positive, lactose-positive
E. coli 0157	Trypticase soy broth containing bile salts dipotassium phosphate and novobiocin	Sorbitol–MacConkey	Sorbitol-negative
Salmonella spp.	Selenite F or Rappaport–Vassiliadis	Desoxycholate citrate agar	Lactose-negative, H_2S production
V. parahaemolyticus	Glucose salt teepol broth	Thiosulphate citrate bile salts sucrose agar	Growth at 42°C, oxidase-positive
Y. enterocolitica	Nutrient broth 'cold enrichment' at 4°C	MacConkey at 25°C	Urease-positive, motile at 22°C

7.2.2 Identification

(a) Bacteria

A number of different characteristics may be useful in rapidly making a presumptive and later a final identification. These include: colonial morphology, staining reactions, growth requirements (such as gaseous conditions and optimal temperature), tests of biochemical activity, and antigenic structure necessary for serological characterization.

Morphology and staining
The Gram stain is an essential first step in the identification of bacteria. The examination of either a stained or unstained film may then provide information on the shape, size and arrangement of bacterial cells. Spore and capsule formation may also be observed, usually by specific staining.

Growth characteristics
Since different bacteria have different nutritional requirements and produce a typical colonial morphology on certain media, we are able to use this information to aid recognition. The ability of an organism to grow at an elevated temperature (like campylobacter) or a low temperature (like listeria)

can be a simple way of providing a clue as to its identity. Similarly, specific gaseous requirements will immediately determine whether the bacterium belongs to the obligately aerobic, facultatively anaerobic, obligately anaerobic or micro-aerophilic category.

Biochemical activity
When a bacterium has been isolated in pure culture, it is then possible to assess its metabolic or biochemical activity by using a range of simple tests. These usually involve the utilization of a substrate, e.g. lactose, which may result in acid production; the formation of specific metabolites, e.g. indole or H_2S; or the breakdown of organic macromolecules, e.g. lecithin.

Serological tests
Tests for antigenic differences fall into two categories: those for surface antigens and those for toxins.

Tests for surface antigens are discussed in more detail in section 7.2.3, but the following example gives an idea of their usefulness in identification. A sorbitol non-fermenting *E. coli* could be positively identified as serotype 0157 using a specific antiserum which only agglutinates *E. coli* 0157 cells.

As regards toxins, sera containing antitoxins for *Cl. botulinum* or *Cl. perfringens* are often incorporated into culture media so that typical colonies can be identified as toxin producers. For example, in the Nagler reaction, a test used to detect *Cl. perfringens*, which produces a lecithinase (the α-toxin) when the organism is grown on egg-yolk agar, opalescence develops around colonies due to the digestion of lecithin. As some other clostridia also produce a lecithinase the reaction is made more specific by spreading *Cl. perfringens* anti-toxin on one half of the plate. Following incubation, although growth occurs in both halves of the plate, strains which produce α-toxin produce opalescence only in that half of the plate which does not contain antitoxin.

(b) Viruses

For those viruses implicated in food poisoning, identification depends primarily on the direct detection of the virus or antigen in clinical material. This is partly because of the rapidity of the techniques involved, and partly because tissue culture for these viruses has been notoriously difficult. Electron microscopy is perhaps the most obvious method whereby virus particles can be detected and provisionally identified on the basis of their characteristic morphology. Serological techniques such as enzyme-linked immunosorbent assays (ELISAs) which have been made more specific by the use of monoclonal antiviral antibodies are used in antigen detection. Similarly, rotaviruses can be detected in stools using latex co-agglutination, where a specific antibody is tagged onto latex particles which agglutinate in the presence of rotavirus antigen.

(c) Protozoa

Faecal samples need to be treated by concentration methods to increase the normally low numbers of cysts present, and then examined for cysts in giardiasis and for oocysts in cryptosporidiosis. Although giardia cysts can be seen in wet preparations using light microscopy, oocysts of cryptosporidia need to be stained either with auramine, which they retain after staining, or with a modified Ziehl–Neelsen method. In the acute phase of giardiasis motile trophozoites may also be seen.

(d) Mycotoxic fungi

Although selective and diagnostic media are widely used in food bacteriology, a comparable methodology is not very well developed for food mycology. A medium containing dichloran, glycerol and chloramphenicol is very useful for the general enumeration of moulds in the presence of bacteria. In addition, *Aspergillus flavus* and *parasiticus* agar (AFPA) can be used for the selective enumeration of these two important species, which produce colonies with a characteristic orange reverse on this medium.

7.2.3 Epidemiological typing

In the investigation of food poisoning outbreaks it is important to be able to type bacteria, that is, to determine whether different isolates of the same genus and species also belong to the same strain. Traditionally, microbiological typing methods have been: biochemical profiling or biotyping, serotyping and bacteriophage typing.

Biotyping by definition means primarily distinguishing types by biochemical and metabolic activities. Biotyping of those bacteria that cause food poisoning is not particularly useful, as many lack unusual biochemical characteristics and some may not be very active biochemically. In addition, many bacteria that can now be more finely discriminated using newer techniques still display the same biotype.

Serotyping is the characterization of a number of antigenically-distinguishable members of a single bacterial species; serologically, bacterial strains may exhibit differences which are not apparent from the results of biochemical tests. One classical example where serotyping has been essential in epidemiology and the monitoring of disease is salmonellosis. With salmonella, the subdivision of the genus is based mainly upon antigenic analysis. More than 2000 salmonella types can be separated on the basis of somatic (O) and flagella (H) antigens, using the Kauffmann and White scheme (section 2.1). This fine differentiation of salmonellae increases the precision and certainty with which the sources and methods of spread of infection can be identified.

Phage typing is a method based on differences in the susceptibility of bacterial strains to a range of bacteriophages. Each of the phages used for typing is lytic for one, or a limited number, of the strains of the species under test. A phage which lyses the particular strain being tested will form a clear macroscopic area against an opaque background layer of surface bacterial growth. The strain can then be defined and identified in terms of the phages to which it is sensitive.

7.2.4 Bacterial toxin detection

Demonstrating the presence of an enterotoxin in food and/or clinical material has advantages of speed and specificity when compared with routine culture. Typically, toxin assays cannot be performed directly and procedures for extraction and/or concentration of enterotoxin need to be performed before assay.

(a) *Cl. botulinum*

Of greatest importance in the diagnosis of botulism is the identification of botulinum neurotoxin in suspected food, in the contents of the alimentary tract, and in the patient's blood. Since botulinum toxins are recognized by their lethal action, it is not surprising that the bioassay remains the most sensitive test for their detection. In addition, the mouse bioassay is highly specific and reliable. However, this bioassay has two disadvantages: protein shock may kill the animal, or it may take up to 4 days to show signs of symptoms when toxin levels are low. Because of this, a number of immunological methods have become available, but only the ELISA is a workable alternative. Unfortunately, the ELISA also has disadvantages: it is less sensitive than the bioassay, and involves the expensive process of preparing and purifying specific antibodies to all toxin types.

(b) *Cl. perfringens*

Although microbiological and serological techniques have been used, the detection of enterotoxin in faeces is the single most important factor in confirming *Cl. perfringens* as the specific cause of food poisoning outbreaks. This has been exemplified in diagnosing the cause of gastroenteritis in elderly patients. In hospitals and homes for the elderly, it may often be normal for symptomless patients to harbour large numbers of spores in the intestinal tract. This would be problematic if the identification of spores in faeces were the only method used to establish *Cl. perfringens* enterotoxin, but they are not as sensitive as more modern serological tests such as ELISA and RPLA (reversed passive latex agglutination). RPLA uses sensitized (antiserum to enterotoxin) latex beads which are exposed to serial dilutions of enterotoxin-containing material. After incubation for 20–24 hours, agglutination is interpreted as a positive result. Because of the simplicity and speed of the RPLA procedure, this is probably the method of choice for enterotoxin detection, although the test kit is relatively expensive. For those laboratories with tissue culture facilities, a Vero cell tissue culture assay for enterotoxin detection is relatively easy. However, this assay is liable to false-positive reactions due to the presence of other toxic or interfering substances in human faeces.

(c) *Staph. aureus*

Until recently, the methods available to most laboratories for the detection of *Staph. aureus* enterotoxin in foods involved a rather lengthy extraction and concentration process, and it was difficult to detect low levels of toxin consistently. Presently, all of the methods for detecting these enterotoxins are based on the use of specific antibodies to each enterotoxin. The most common methods used are ELISA and RPLA, which are sensitive and now available commercially.

(d) *B. cereus*

It is possible to detect the diarrhoeal type of *B. cereus* enterotoxin in a variety of foods with a range of *in vivo* and *in vitro* assays. Although the *in vivo* assays might be useful in understanding pathogenesis, they are not particularly suitable for laboratory diagnosis. Nowadays, we have an immunological *in vitro* assay such as RPLA, which is simple, rapid and shows good sensitivity. At present, there are no *in vitro* assays available for the emetic toxin.

7.2.5 Detection of mycotoxins

As the mycotoxins are a chemically diverse group of relatively small molecular weight metabolities, they have presented problems for analysis. Traditionally, a typical analytical procedure for their detection might involve sample preparation and extraction, a clean-up stage followed by concentration, and finally analysis using a chromatographic procedure such as thin-layer chromatography (TLC), high-pressure liquid chromatography (HPLC) or gas-liquid chromatography (GLC). Obviously, the workload and time involved would be an enormous burden in the routine monitoring of foods for the presence of mycotoxins. More recently, immunological techniques have become available as a result of the raising of specific antibodies, made possible by coupling the toxins to suitable carrier macromolecules. Perhaps the most acceptable of these newer methods have been the direct and indirect ELISA techniques which have been developed for the aflatoxins, zearalenone, ochratoxin and T-2 toxin, which show good specificity and sensitivity, and which have now been made commercially available for aflatoxins.

One disadvantage of some of these techniques is that for some bacteria neither serotyping nor phage-typing systems exist. A second is that occasionally, an outbreak of a disease is caused by an organism with commonly occurring phenotypic characteristics. Other problems may include autoagglutination (a source of false-positive results) in serotyping, and the need to select precise environmental conditions for optimal toxin elaboration. These drawbacks have prompted the development of new and simple genetic techniques known as 'molecular epidemiology' which are described in section 7.3.5.

7.3 NEWER METHODS

The most appropriate treatment of patients is only possible when we achieve definitive rather than circumstantial evidence for an association between the ingestion of a particular food and disease. The use of conventional techniques for detecting food poisoning organisms has placed constraints on this process in terms of the time needed to make a laboratory diagnosis. From the information contained in Chapters 2, 3, 4, 5 and 6 it is obvious that we need to be able to detect these organisms as accurately and as rapidly as possible during outbreaks of food poisoning. Some exciting new methods for the rapid detection of micro-organisms will be briefly described and discussed below.

7.3.1 Media and cultural techniques

Most of the newer developments in culture media have been in the creation of different selective and/or differential media formulations to increase the isolation of a pathogen from a particular food. The effect has been to optimize the efficiency of recovery from slightly more hostile environments and so broaden the range of foodstuffs that can be examined successfully. A number of newer media also now allow us to make a presumptive identification of pathogens on selective/differential media and so enable a final identification to be made more rapidly.

Recently, a number of commercially available culture test kits have been introduced which facilitate more rapid detection of certain pathogens. One of these test kits utilizes a selective enrichment motility technique for detecting salmonellae in foods. Following pre-enrichment, a tube containing a selective medium and an upper indicator medium, separated by a porous partition, is inoculated. Any salmonellae present will migrate through the lower selective medium to the upper indicator medium, where their presence is denoted by a colour change. Other kits concentrate organisms on a filter; when a specific antiserum is then added, an immunodiffusion band is formed at the interface.

7.3.2 Immunoassays

The most widely used immunoassay is the ELISA. This technique is used for detecting and quantifying specific serum antibodies or microbial antigens (Figure 7.1). Specificity is increased in antigen detection if monoclonal antibodies are used. These are a population of identical antibodies, all of which recognize the same specific determinant on a simple or complex antigen.

The production of genus-specific monoclonal antibodies for listeria and salmonella has allowed the development of ELISA tests for these bacteria. These tests can be used directly on broth samples (usually after selective enrichment), without the need for pure cultures, and have provided presumptive results 1 day earlier than with standard cultural methods. The ELISA tests for salmonella show specificity and sensitivity levels which are as good as conventional culture and serotyping methods. Unfortunately, ELISA tests for listeria detect several non-pathogenic strains as well as *L. monocytogenes*, and cannot discriminate between them. This can be a problem, as these other species of listeria often occur as innocuous saprophytes in the environment.

Other applications of the ELISA technique have been in toxin determination and in the diagnosis of protozoal infections. Monoclonal antibodies have been produced for the heat-stable (ST) and heat-labile (LT) toxins of *E. coli*, and for *Cl. botulinum* toxin. For the latter a monoclonal antibody was selected for its high binding capacity to botulinum (type A) toxin, resulting in a highly sensitive test which shows good promise.

Although not yet fully developed, an ELISA test has been produced to detect *G. lamblia* antigen in human faecal samples, with good results. Finally, an ELISA test has been devised to help in the diagnosis of cryptosporidiosis, by establishing the presence of specific IgM and IgG antibodies.

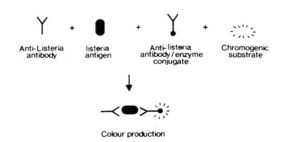

Figure 7.1   Basis of the ELISA test. (Source: Eley, A. (1990) *British Food Journal*, **92**, 28–31.)

7.3.3 Immunomagnetic separation (IMS)

Immunomagnetic separation using magnetic beads coated with antibodies specific to surface antigens of bacteria, has been shown to be efficient in isolating different bacteria from food and clinical samples (Figure 7.2).

MAGNETIC BEADS

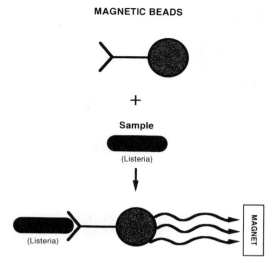

Figure 7.2 Immunomagnetic capture of listeria on microscopic magnetic beads coated with antibodies directed against listeria.

Indeed, the time taken for testing can be shortened and enrichment eliminated by isolating bacteria such as salmonella, *Staph. aureus* and listeria directly from food samples. It has been shown that immunomagnetic techniques can be used to replace the classical 24-hour microbial selective or non-selective enrichment steps by a 10-minute procedure. Like other tests it is important to consider different parameters for its optimization, such as the concentration of beads, incubation time, mode of incubation, type of sample and the washing regimes employed. It is also known that use of too high a concentration of beads and too long an incubation time increases non-specific binding of competitive flora, while the use of a sub-optimum concentration of beads reduces sensitivity.

Recently, in a study of spiked environmental samples, detection of listeria by capture on antibody-coated magnetic beads has been shown to decrease test time and improve sensitivity, relative to cultural methods. In this study, immunomagnetic capture was compared with standard cultural methods for detection of listeria in a broad range of spiked and naturally contaminated food and environmental samples. Immunomagnetic capture was at least as sensitive as cultural methods for detection of listeria in seafood, meats, dairy foods, and environmental samples. It was possible to determine the number of listeria present in a sample, because immunomagnetic capture was carried out directly from the sample, without enrichment. These quantitative results were produced within 24 hours while cultural methods required 6–14 days to produce a qualitative result. Immunomagnetic capture was thus more rapid and as sensitive as standard cultural methods for detection of listeria in the food and environmental samples tested.

7.3.4 Nucleic acid probes

Recent advances in molecular biology, including the use of nucleic acid probes, have led to the creation of novel methods of detecting pathogenic micro-organisms from food samples and clinical specimens.

Probes may be defined as small, labelled (i.e. detectable) pieces of nucleic acid which can seek out and bind to fragments of target DNA or RNA (in the food or clinical sample) with complementary sequences. This process, known as hybridization, forms the basis of this technology (Figure 7.3).

Nucleic acid probes are now in use or under development for all the major food poisoning bacteria and are usually directed at known virulence determinants to increase specificity. This is because not all strains of a particular species are pathogenic to humans. Using conventional methods, it is not always a straightforward matter to assess the potential virulence for humans of a strain in the laboratory.

In many respects, probe technology represents a significant advance over conventional techniques in the detection of micro-organisms. One major advantage of the use of probes in the diagnostic laboratory has been a reduction in the time needed for the identification of fastidious pathogens, including their direct detection in clinical specimens. Because of the nature of DNA hybridiza-

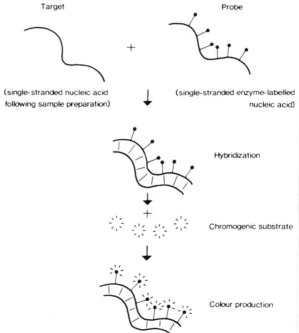

Figure 7.3   Basis of the non-isotopic nucleic acid probe test. (Source: Eley, A. (1990) *British Food Journal*, **92**, 28–31.)

tion reactions, the resultant DNA hybrid should be highly specific, and the technique is not growth dependent, which is useful for demanding or slow-growing organisms, such as some obligate anaerobes. Commercially produced probe kits for routine use are generally easy to use and allow a broader range of organisms to be identified in smaller laboratories.

However, although some probe tests can be completed in less than 1 hour, many will take from 1–2 days from the time the specimen arrives in the laboratory to completion, because of the enrichment phase required for some pathogens. These particular tests cannot, therefore, be described as rapid procedures. Moreover, despite attempts to increase the sensitivity of probe tests, we are still a long way from achieving results which are comparable with those of bacterial culture, which in theory requires only one viable cell for growth. Furthermore, at present nucleic acid probe technology is relatively expensive.

A solution to the current problem of poor sensitivity of probes now lies in amplification technology such as the use of the polymerase chain reaction (PCR). PCR was first described in 1985 and has significantly transformed the usefulness of nucleic acid probe tests. In a nutshell, PCR is a method of amplifying a target DNA sequence *in vitro*, typically by a factor of 100 000 or more in a few hours. The result is that with this technique the limitations of poor sensitivity of conventional DNA probe technology have now been greatly reduced, and the time-consuming enrichment process has been eliminated. However, it does need to be stressed that the target nucleic acid sequence from the sample needs to be known in order to construct the primers accordingly.

The PCR technique takes advantage of the natural DNA replication that occurs when cells divide. During this process the two strands of DNA separate and copies are made to enable a full set of genes to be passed for each new cell, so that after each replication cycle there is a doubling of the number of copies of the original DNA. The principle of PCR is simple, requiring a three-step cycling process:

1. denaturation of double-stranded DNA;
2. annealing of primers; short pieces of DNA with sequences complementary to the DNA on either side of the target sequence; and
3. primer extension.

Denaturation separates the complementary strands of DNA held together in the duplex by hydrogen bonds. Although there are several physical and chemical means of dissociating the duplex, heating it to 95–100°C is simple and efficient. In the annealing process (approx. 40°C), primers are attached to the dissociated DNA strands. Each primer is complementary to one of the original DNA strands, to either the left or right of the sequences of interest. Once annealing has occurred, an enzyme catalyses the synthesis of new strands of DNA. The enzyme is a DNA polymerase that adds nucleotides (from nucleotide triphosphates) complementary to those in the unpaired DNA strand onto the annealed primer

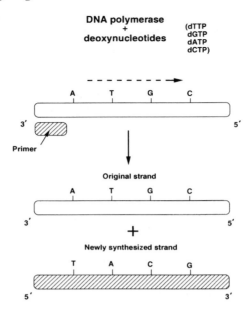

Figure 7.4   Primer extension and the generation of a new strand of DNA. (Source: Eley, A. (1993) *Nutrition and Food Science*, **2**, 9–13.)

(primer extension at approx. 70°C) (Figure 7.4). Use of a thermostable DNA polymerase (*Taq* polymerase) eliminates the need to add enzyme after each cycle and allows the amplification technique to be automated. As the number of strands doubles upon completion of each cycle, after 30 cycles a single copy of DNA can be increased up to 1 million copies (Figure 7.5).

A number of different studies using PCR have been reported including those on campylobacter, shigella, vibrio, staphylococci and salmonella, etc. In almost all cases there has been a significant reduction in time needed to produce a result, often with good levels of sensitivity.

7.3.5 On-line monitoring

More recently, newer techniques have emerged that show potential for on-line monitoring such as flow cytometry and lux recombinant bacteriophage.

(a) Flow cytometry

The principle behind this technique is that when a food sample (to be used as a cell suspension) flows in a continuous stream though a focused laser beam, optical signals are detected whenever a microscopic particle is encountered. This technique is powerful, as it can analyse particles at rates of up to 10 000 per second, and for each particle a number of different properties such as size,

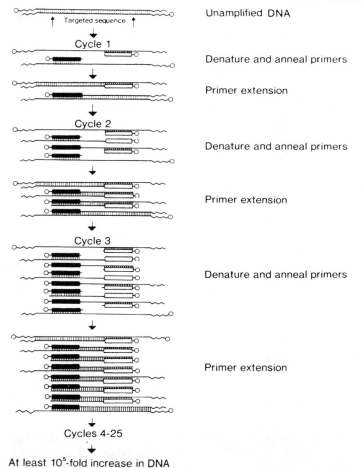

Unamplified DNA

Targeted sequence

Cycle 1

Denature and anneal primers

Primer extension

Cycle 2

Denature and anneal primers

Primer extension

Cycle 3

Denature and anneal primers

Primer extension

Cycles 4-25

At least 10^5-fold increase in DNA

Figure 7.5 The Gene Amp™ Polymerase Chain Reaction. (Gene-AMP™ is a trademark of Cetus Corporation). (Source: Eley, A., in Borriello, S.P. (ed.) (1990) *Clinical and Molecular Aspects of Anaerobes*, pp. 275–83.)

shape and DNA content can be measured. Although this technique is still in its experimental stage for microbiological analysis, the use of one or two parameters (size, and nucleic acid content) can identify different bacterial species at high levels of sensitivity. At present the machinery for flow cytometry is expensive and complicated. However, it has enormous potential for on-line monitoring of food safety in factories, as it will allow the direct detection and identification of bacteria from food within minutes.

(b) Lux recombinant bacteriophage

Rather than look for pathogens directly, in some circumstances it might be useful to detect indicator micro-organisms. By definition the latter are organisms present in significant numbers within a food which, while not pathogenic, can be related through increasing count, to the increased probability of pathogen contamination. The principle of this test requires the introduction of the lux genes encoding bacterial luciferase into the genome of a bacteriophage. The recombinant phage lack the intracellular biochemistry necessary for light production and, in consequence are dark. Infection of host bacteria by the phage, however, leads to the expression of host phage genes and within 30–50 minutes, the additional lux genes. The result of phage infection is bioluminescent bacteria and it is possible to harness the biological specificity of bacteriophage to confer a bioluminescent phenotype on a defined set of bacteria, growing in a complex microbial mixture of organisms. Recombinant lux bacteriophages (e.g. P22) have been used to detect enteric indicator bacteria without recovery or enrichment in 50 minutes, provided they are present at levels greater than 10^4 per g and allow the detection of as few as 100 bacterial cells. As P22 is a narrow-range bacteriophage infecting only *S. typhimurium*, its lux-containing derivative provides a specific test for this organism, even if it is present as only one cell in a million in a bacterial mixture. It should be stressed that this technique is still in the experimental stage, but it has the potential to become a rapid, sensitive and specific assay for a broad range of bacteria, including all those that cause food poisoning.

7.3.6 Epidemiological typing

Although we have described conventional techniques for epidemiological typing of bacteria, these methods suffer from one or more of the following disadvantages:

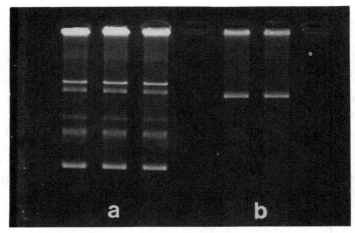

Figure 7.6 Plasmid analysis of five bacterial isolates showing two groups of strains: (a) containing five plasmids; (b) containing one plasmid .

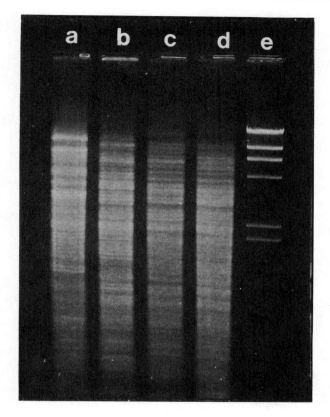

Figure 7.7 Chromosomal DNA fingerprinting of four bacterial isolates showing two dissimilar (a,b) and two identical (c,d) strains. e is a DNA marker.

1. they rely on phenotypic characteristics that may not be stably expressed;
2. the necessary reagents may not be commercially available;
3. they may not be sensitive enough to distinguish each strain of a species;
4. the system may be applicable to only one bacterial species.

Recent developments in DNA analysis techniques have reduced the dependence on detecting phenotypes and have allowed us to look at the 'molecular epidemiology' of bacteria. The three main approaches involve plasmid DNA analysis, chromosomal DNA analysis and nucleic acid probes.

Plasmids are small, circular forms of DNA which in general exist independently of the bacterial chromosome. Most plasmids are relatively stable and can now be easily isolated. Plasmid analysis or 'fingerprinting' (with or without the application of restriction endonuclease cleavage) has been used to characterize and separate bacterial strains that possess the same bacteriophage type or serotype (Figure 7.6).

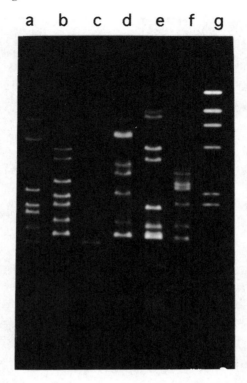

Figure 7.8 Use of an rRNA probe to show dissimilar bands after hybridization with bacterial isolates (a–f). g is a DNA marker.

The second approach achieves the same end by comparing the banding patterns of DNA fragments in an agarose gel which result from restriction endonuclease cleavage of extracted chromosomal DNA (Figure 7.7). However, the usefulness of such patterns as diagnostic tools is limited by their complexity: they may be made up of hundreds of bands. It is difficult to identify minor but possibly significant changes in banding patterns, and to reproduce identical bands on a day-to-day basis. This technique can be applied more successfully to viral RNA (electropherotyping), and has been widely used as a tool for studying the molecular epidemiology of rotaviruses in the community.

Since there are problems with chromosomal DNA analysis, it is now possible to use nucleic acid probes to highlight specific DNA restriction site heterogeneities and so reduce the number of bands for analysis (Figure 7.8). Reducing the number of bands enhances the feasibility of developing a highly discriminatory genetic typing system, applicable to many isolates. This technology provides a valuable new method of distinguishing among related species of micro-organisms, and of detecting variation among strains within species. There have now been many applications of these three typing methods to food poisoning organisms and more recently several newer techniques have emerged; these include pulsed field gel electrophoresis, multilocus enzyme electrophoresis and PCR ribotyping. There can be no doubt that with the increased use and development of molecular methods, there is great potential for the precise identification of the microbial causes of outbreaks in the future.

8 Epidemiology

J.C.M. Sharp

8.1 TRENDS IN FOOD POISONING

In the UK, a marked increase has been observed in the reported incidence of food poisoning since the mid-1960s, a phenomenon also seen in most other countries of Europe and in North America. Notifications of food poisoning in the UK more than quadrupled from 17 300 in 1982 to over 74 000 in 1993. Reports of laboratory diagnosed infections such as salmonellosis, campylobacter enteritis and haemorrhagic colitis due to verotoxin-producing strains of *Escherichia coli* (VTEC), increased between two- and 40-fold over the same period. To what extent these increases were real or due to improved ascertainment, to more comprehensive investigation and better reporting and/or to enhanced surveillance, is uncertain. Nevertheless, a significant increase has been observed worldwide in the reported incidence of food poisoning.

In contrast, the more traditional forms of food poisoning of the 1960s and 1970s, caused by enterotoxin-producing strains of *Staphylococcus aureus* and heat-resistant *Clostridium perfringens*, have declined in importance. Meanwhile other infections of bacterial (e.g. *Aeromonas* spp., *Vibrio* spp., *Yersinia* spp.), viral (e.g. small round structured viruses [SRSV], small round viruses [SRV]) and protozoal (e.g. *Cryptosporidum, Giardia*) aetiology, have increasingly become associated with food- or water-borne spread.

Food poisoning in developed countries of the world in recent decades may conveniently be divided into three broad categories, viz. those of increased importance (i.e. in their reported frequency and/or clinical severity), those of decreased importance and those where the role of food-borne transmission is not yet clear.

Increased importance

Botulism	Listeriosis
Campylobacter enteritis	Salmonellosis
E. coli (VTEC)	Viral gastroenteritis
haemorrhagic colitis	(SRSVs, SRVs, etc.)

Decreased importance

Bacillus spp. (*B. cereus*, etc.)	Staphylococcal enterotoxin
Clostridium perfringens	*E. coli* gastro-enteritis (other than VTEC)

Uncertain significance

Aeromonas spp.	*Streptococcus zooepidemicus*
Cryptosporidiosis	Vibriosis (*V. parahaemolyticus*, etc.)
Giardiasis	Yersiniosis (*Y. enterocolitica*, etc.)

With the exception of food-borne viruses, those organisms of increased import-ance are primarily associated with foods of animal origin, viz. meats or dairy products which become contaminated during milking or post-slaughter pro-cessing from the faeces of infected carrier animals (*Salmonella* spp., *Campylo-bacter* spp., VTEC), or indirectly from the environment during food production (*Clostridium botulinum*, *Listeria monocytogenes*).

In contrast, the transmission of *Shigella* spp. and *E. coli*, etc. which are primarily of human origin, via contaminated food has become a rare but not unknown occurrence in most developed countries. For example, in the early summer of 1994 an international outbreak of *Sh. sonnei* dysentery associated with imported iceberg lettuce from southern Europe, affected an unquantifiable number of persons in Sweden, Norway and the UK. However, indirect trans-mission of enteropathogens of human origin via sewage-contaminated waters and the consumption of shellfish, in particular uncooked oysters, is a major cause of viral gastroenteritis.

When they occur, episodes of food poisoning may present in the form of:

1. Sporadic cases with no apparent association with any other cases;
2. Family outbreaks affecting two or more persons within a household; or
3. General community outbreaks affecting groups of persons who have eaten in the same restaurants or canteens, within residential institutions (hos-pitals, old persons' homes, prisons, etc.), or who have eaten contaminated foods obtained from the same retail (or wholesale) outlet(s).

8.2 FACTORS CONTRIBUTING TO FOOD POISONING

A review of over 500 outbreaks of salmonellosis in England during the 1970s and early 1980s, showed that the main contributory factors, several of which coexisted, were:

Preparation too far in advance	42%

Storage at ambient room temperature	30%
Undercooking	25%
Inadequate cooling	22%
Cross-contamination	15%
Inadequate re-heating	13%
Inadequate thawing	11%
Use of left-overs	4%

Similar factors contribute to a greater or lesser extent in other forms of food poisoning, with poor temperature control compounding all other contributory factors.

While faulty practices during food preparation frequently contribute to the growth of food poisoning organisms and the spread of contamination within a kitchen, in only 2% of outbreaks was an infected food-handler thought to have been responsible. The infected food-handler, however, is usually the main source in outbreaks of staphylococcal food poisoning and viral infections.

8.3 EPIDEMIOLOGICAL FEATURES

Considerable variation may exist in the frequency of food poisoning in relation to the age of those affected, the time of year and the location of outbreaks.

8.3.1 Age incidence

Food poisoning can affect both sexes more or less equally, with all ages potentially being at risk. Nevertheless, the highest reported incidence, particularly where confirmed by the laboratory, occurs among infants and pre-school children. These are the age groups along with the elderly, who are more likely to be seriously ill, and/or to have consulted their family doctor.

8.3.2 Seasonal incidence

Food poisoning generally occurs most frequently over the warmer months of the year, with salmonellosis in particular showing a preponderance of cases during the summer and early autumn (Figure 8.1). Infections due to *Campylobacter* spp. and viruses, which are unable to grow on food and are less dependent on warmer ambient temperatures, differ in their seasonal patterns. Campylobacter infections usually peak during the late spring and early summer some 6–8 weeks before the salmonellas, followed by a secondary autumn rise. Viral infections associated with seafoods feature most frequently over the colder winter months.

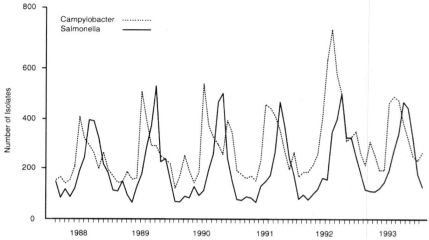

Figure 8.1   Seasonal distribution of *Campylobacter spp.* and *Salmonella* spp. isolates reported to SCIEH by laboratories in Scotland by 4-week periods, 1988–1993.

8.3.3 Location of outbreaks

Commercial catering establishments (hotels, restaurants and canteens) and residential institutions (old peoples' homes, prisons, etc.) feature most frequently as the venue where food was contaminated or mishandled in general outbreaks. Family outbreaks are usually a consequence of faulty food preparation practices (inadequate cooking, cross-contamination, etc.) within the domestic kitchen.

8.4 THE FOOD INDUSTRY

Eating habits and food production methods have changed dramatically in industralized countries as a result of the many social and economic changes which have taken place since the end of the Second World War. More people eat communally in snack bars, canteens and restaurants, while increasing demands are made for fast foods, convenience foods (pre-cooked, frozen, chilled, etc.) and for better, cheaper and 'healthier' foods.

The farming and food industries have successfully evolved in such a way that most developed countries are now largely self-sufficient in food production. In the more affluent societies, consumer demands are continually being made for improvements in standards and types of food produced. However, moves towards less-intensive organic production methods (with fewer added preservatives, etc.) may be accompanied by increased risks of exposure to microbial contamination.

The requirements thus placed upon modern farming impose considerable economic pressures to improve efficiency by greater stocking densities. Dairy herd sizes in the UK, for example, have doubled since the 1960s without any corresponding increase in staff. As a result the number of cows being milked per dairyman has increased, with less time being available to attend to the welfare or hygiene of individual animals. Increased livestock densities (seen particularly in the poultry-broiler industry), complemented by a more rapid turnover of animals (or birds), results in greater opportunity for cross-infection to occur, and consequently enhances the likelihood of cross-contamination of carcasses during post-slaughter processing.

As a result of the changing and widening patterns in national and international distribution of foods, outbreaks of food poisoning have nowadays become much more widespread in their geographical presentation. Simultaneously, the development of the cold-chain in food distribution, along with the increased use of the commercial deep-freeze may also distort the time-scale of the outbreaks. Hence selected cases may appear at different times and in different parts of the country (or countries), and outbreaks are thereby frequently less well-defined or recognizable as such than when confined to a localized incident.

Foods which have been heat-treated by pasteurization (milk, ice-cream) or by sterilization (canned foods, etc.) are generally safe. Other foods (preserves, fruits, fats, flour, etc.) are equally regarded as being safe as a consequence of their pH levels, water, salt and/or sugar content providing conditions unsuitable for bacterial growth and toxin production.

8.4.1 The poultry-broiler industry

The potential dangers of cross-infection and cross-contamination are greatest within the poultry-broiler industry, where highly intensive systems of breeding, rearing and processing (factory farming) have evolved. In 1993, over 700 million chickens and turkeys were produced for human consumption in the UK. Broiler chickens are not uncommonly reared in sheds containing up to 20 000 birds.

At the age of 6–7 weeks, when about 5% of a flock may be carrying salmonella organisms in their intestinal tracts, birds are slaughtered at rates which may exceed 10 000 per hour. The stress of transport from rearing farms to the processing plant exacerbates salmonella excretion and cross-infection, while the process of defeathering, evisceration and chilling readily contribute to cross-contamination of carcasses. The waste products (inedible offal, feathers, heads, feet, etc.), which are invaluable sources of protein, are fed back to growing birds and the salmonella cycle is thus maintained. New salmonella strains may be introduced via imported breeding stock or contaminated feed, or less frequently from vermin or the environment. Infection is usually passed vertically from grandparent birds to breeding stock, thence to production flocks and can be transmitted horizontally thereafter within individual flocks (Figure 8.2).

Surveys of retail poultry meat in England between 1979 and 1990 by the Public Health Laboratory Service (PHLS) demonstrated salmonella contamination

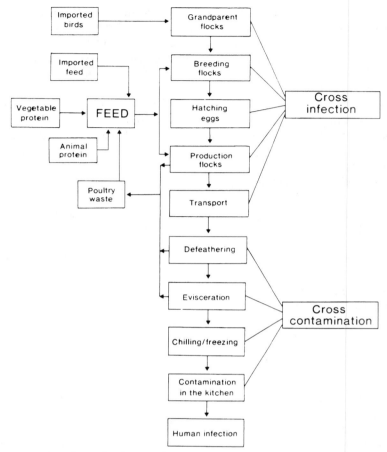

Figure 8.2 The salmonella chain in the poultry broiler industry. After Reilly, W.J. (1991) *Reviews in Medical Microbiology*, **2**, 170–6.

rates varying between 42% and 79% of birds examined. Surveys in other countries have demonstrated similar recovery rates from turkey and chicken carcasses. Subsequent studies have indicated that up to 100% of young broiler birds may be carrying campylobacter organisms in their gut. In one survey in the UK, *C. jejuni* was isolated from 22 of 46 (48%) fresh birds, 12 out of 12 uneviscerated ('New York dressed') birds, and one (4%) of 24 frozen birds sampled. Between 70–80% of chicken carcasses in a processing plant in Israel were shown to be contaminated with *C. jejuni* or *C. coli*.

8.4.2 The dairy industry

Contamination of milk from animal faeces during milking cannot be totally avoided, even in the best-run dairies. Outbreaks of milk-borne infection associated with untreated milk supplies, were a major public health problem in Scotland and to a lesser extent in England and Wales during the 1970s and early 1980s.

In August 1983, legislation (The Milk Special Designations Scotland Order 1980) was introduced in Scotland requiring the heat-treatment of cows' milk for retail sale, following which the large community outbreaks of milk-borne salmonellosis and campylobacter enteritis previously experienced, were effectively controlled. Similar legislation was equally effective when introduced in Canada in 1991. In England and Wales, the Milk Special Designation Amendment Regulations 1985 also banned the sale of untreated cows' milk in shops, supermarkets, canteens and other retail outlets, but continued to permit sales direct from dairy farms in rural areas. Outbreaks of milk-borne infection (salmonellosis, campylobacter enteritis and *E. coli* 0157) continue to be reported from time-to-time. In 1990 more stringent microbiological standards were introduced, with untreated milk for retail sale also being required to carry the warning 'This milk has not been heat-treated and may contain organisms harmful to health'.

Untreated milk can still be sold in many rural districts of most European countries, and under certain licensing conditions, in many states of the USA, although untreated dairy products are generally not widely available. Outbreaks of campylobacteriosis, salmonellosis, listeriosis, *E. coli* 0157 and staphylococcal food poisoning associated with untreated products from cows', goats' and sheep milk have been recorded.

Heat-treatment of milk should be complemented by good hygiene and correct operating practices at all stages between producer and consumer. Outbreaks of milk-borne infection (salmonellosis, campylobacteriosis, listeriosis, staphylococcal food poisoning etc.) associated with contaminated or inadequately heat-treated milk products have been reported in North America, the UK and several other European countries. The most extensive of these, due to *S. typhimurium*, occurred in 1985 in Illinois and adjacent states in the USA and affected an estimated 200 000 persons. A nationwide outbreak (caused by *S. ealing*) over several months in 1985–1986 in widely scattered areas of the UK affected many infants and other persons who had consumed a particular brand of contaminated powdered milk. The importance of proper heat-treatment complemented by hygienic operating practices has been re-iterated on occasion in recent years in Scotland with community outbreaks of salmonella and *E. coli* 0157 associated with inadequately heat-treated milk supplies. Between January and July 1984, over 1500 persons were affected in the Maritime provinces of Canada as a consequence of eating salmonella-contaminated cheddar cheese. Outbreaks of listeriosis attributed to soft cheeses from inadequately heat-treated and raw milk have been reported from the USA and Switzerland.

8.4.3 Mass catering

Special interest has been taken in recent years in the role of mass catering, particularly in hospitals and airlines in relation to outbreaks of food poisoning.

(a) Hospitals

In the 19th century Florence Nightingale commented that 'the first requirement of a hospital is that it should do the sick no harm'. Outbreaks in hospitals and other health-care units are potentially serious given the enhanced vulnerability of patients to food poisoning micro-organisms.

While most hospital kitchens in the UK are of a high standard, it was only with the removal of crown immunity from hospitals in 1987 that a new realism and urgency for improvement in hygiene standards became apparent. Until then, hospitals as crown property had been protected against prosecution from any contraventions of the UK food hygiene regulations.

During the 1970s and 1980s numerous outbreaks were reported, in particular associated with heat-resistant strains of *Cl. perfringens*. Salmonella outbreaks, however, are clinically more serious, with deaths among immuno-compromised patients a not infrequent occurrence; 19 patients died in a psychiatric hospital outbreak in Yorkshire in 1984. In more recent years, however, there has been a marked reduction in reports of hospital outbreaks of food poisoning in the UK, although it is not yet clear to what extent this is attributable to the improvement in catering practices following the removal of crown immunity in 1987. While the hospital may have been over-represented as a venue of food poisoning compared with other forms of communal catering, more comprehensive reporting is expected in that hospitals are semi-closed communities where any outbreak occurring should be recognized at an early stage, in addition to which laboratory facilities for microbiological investigations are more readily available.

(b) Airline catering

Considering the enormous number of in-flight meals prepared and served every day throughout the world, outbreaks of food poisoning associated with airline catering are remarkably rare. Nevertheless, outbreaks of salmonellosis, staphylococcal food poisoning, etc. have been reported from time to time, some of which were both extensive and devastating in their effects. Of these, the largest affected over 760 persons aboard transatlantic flights from London Heathrow in March 1984, following contaminated (*S. enteritidis*) aspic glaze on hors d'oeuvres being served to first-class passengers.

Following this outbreak, a survey of the microbiological quality of airline meals produced by flight catering units serving Heathrow Airport, undertaken by the PHLS Food Hygiene Laboratory, indicated that 24% of 1013 food samples had an excessive surface colony count ($>10^6$ organisms/g) with salmonella organisms detected in four (0.4%) samples. Other organisms detected were *Staph. aureus*, *Cl. perfringens*, *B. cereus* and *E. coli*. Starter (hors d'oeuvres, prawn cocktail, etc.) and main course dishes were equally contaminated.

(c) Home-catering

Domestic kitchens, which are intended for the preparation of food for small family groups, may be used to cater for larger events such as wedding receptions and similar functions. Although food poisoning outbreaks that have occurred have most commonly been associated with a 'one-off' event where catering is for a family party or wedding, others have been in circumstances where a domestic kitchen is used commercially on a regular basis for the preparation of meals. Limitations in the available space for preparation and storage, in particular refrigeration, can pose problems of cross-contamination and temperature-control before, during and after transportation and delivery. Numerous outbreaks of salmonellosis, viral gastroenteritis and *Cl. perfringens* food poisoning in particular have been reported.

8.5 NOTIFICATION AND REPORTING

In England and Wales, a medical practitioner who becomes aware, or suspects, that a patient he or she is attending is suffering from food poisoning, is legally required under the Public Health (Control of Diseases) Act 1984 to notify the 'Proper Officer' of the appropriate local health authority. There is a similar requirement under the Public Health (Notification of Infectious Diseases) (Scotland) Regulations 1988 to notify the Director of Public Health/Chief Administrative Medical Officer of the Health Board of the area, or his representative, and in Northern Ireland under the Public Health (NI) Act 1967 to the Director of Public Health of the appropriate Health and Social Services Board. Laboratory confirmation is not required, although this may be available in many cases. Notification of food poisoning may also be made by a member of the public. Other countries have similar, but variable, statutory requirements covering a wide range of food-related infections.

It is essential that notification of suspected food poisoning is made as early as possible, preferably by telephone, if a meaningful epidemiological investigation is to be made. Early reporting (by the family doctor or by the laboratory) of episodes of specific infections which may not be statutorily notifiable (e.g. campylobacteriosis, listeriosis, VTEC, viral gastroenteritis, etc.), is equally important (Figure 8.3).

The notification process is used not only to alert the local public health authorities so that investigation and preventive action can be undertaken, but also to provide epidemiological data for surveillance purposes. Data relating to notified cases of food poisoning (along with other notifiable infections) in England and Wales are forwarded to the Office of Population Censuses & Surveys (OPCS) in London, and in Scotland to the Information and Statistics Division (ISD) of the Common Services Agency of the Scottish Health Service,

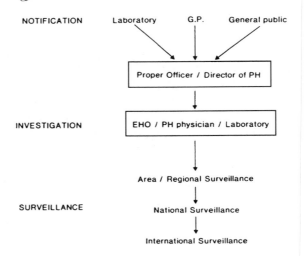

Figure 8.3 Notification, investigation and surveillance of food poisoning.

Edinburgh, for analysis and publication on a weekly and annual basis. In Northern Ireland, notification data are forwarded to the Department of Health and Social Security (NI) and are published 4-weekly and annually.

Over the 20 years up until 1985, a steady increase had been observed in the number of notifications of food poisoning made in the UK, following which a four-fold rise occurred with over 74 000 cases being recorded in 1993 (Figure 8.4).

While food poisoning has been statutorily notifiable since the late 1930s, the more meaningful epidemiological data obtained in recent years are based on voluntary laboratory reporting of specific food-borne pathogens (*Salmonella*

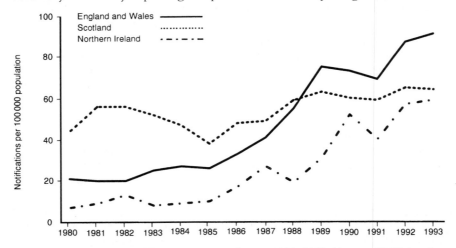

Figure 8.4 Notifications of food poisoning in the UK, 1980–1993. (Source: OPCS, London and ISD, Edinburgh.)

spp., *Campylobacter* spp., VTEC, etc.) complemented by surveillance programmes and epidemiological information obtained by outbreak investigation. It is estimated, however, that notifications of food poisoning and laboratory reports probably represent only the 'tip of the iceberg' and that less than 10% of the real incidence of infection occurring in the community is officially recorded (Figure 8.5).

This is thought particularly to apply to infections caused by non-typhoid salmonellae and campylobacters which are often self-limiting and consequently do not come to the attention of medical practitioners. In addition, the family doctor may omit to notify the local health authority and/or arrange for a faecal specimen from the patient(s) to be sent for laboratory examination. Viral infections are even less likely to be identified. Not only is the illness usually of relatively short duration, with the patient having recovered within 24–48 hours, but even when a viral diagnosis is considered, it is essential that faecal specimens are sent to the laboratory within 24–48 hours of onset, as viruses are rapidly cleared from the stool within 1–2 days. The failure to identify a possible viral

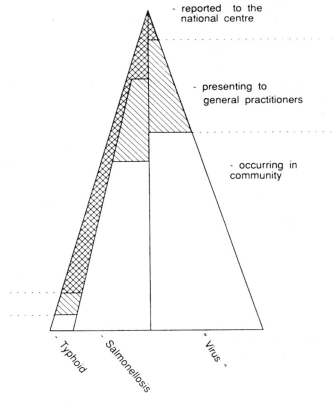

Figure 8.5 Reported and unreported gastrointestinal infection. After Cowden, J.M. (1991; personal communication).

aetiology may be compounded by the fact that electron microscopy examination may not be requested or facilities may not be readily available.

8.6 INVESTIGATION

The successful investigation of any episode of food poisoning is greatly enhanced by the prompt notification of the incident by the medical practitioner in charge of the patient, or by the laboratory which has detected the relevant pathogen (Figure 8.3).

Following notification or a laboratory report, a visit is usually made to the home of the affected person(s) by an environmental health officer (EHO) to initiate the epidemiological investigation. In some areas of the country this may be undertaken by a public health nurse. The significance of a single sporadic case of food poisoning occurring within a practice or district must never be minimized or ignored.

This could be the first indication of a general community outbreak affecting geographically scattered persons who had eaten at the same restaurant, attended the same social function or had purchased a contaminated food from different retail outlets across the country.

In an outbreak occurring in a hospital or health care unit, the investigation is undertaken by the Control of Infection team, which usually includes a microbiologist, a clinician, a control of infection nurse, a public health physician and possibly also an EHO. In larger, community outbreaks, particularly if they are multi-district in extent, the Communicable Disease Surveillance Centre (CDSC), CDSC Wales, the Scottish Centre for Infection and Environmental Health (SCIEH) or the DHSS (NI) as appropriate should be advised at an early stage and invited to participate in the investigations.

In every food poisoning incident, whether in hospital or in the outside community, a close working relationship between the laboratory staff, the public health physician, the EHO, and where relevant veterinary colleagues, is of paramount importance.

In any investigation, the cardinal features which should always be noted, are:

1. **Person** – who was affected? What common features were present? Were they mainly infants, pre-school children attending a day-care centre, employees of the same firm or members of the same golf club?
2. **Place** – when did the incident occur? Was the outbreak associated with a particular hotel or canteen, take-away food from the same restaurant, a wedding party or other social function?
3. **Time** – when did the illness start? Did all those affected fall ill within a few hours of each other, or were they spread over several days or more?

The plotting of an epidemic curve of the dates of onset of those affected, along with other epidemiological characteristics, will usually give an indication as to

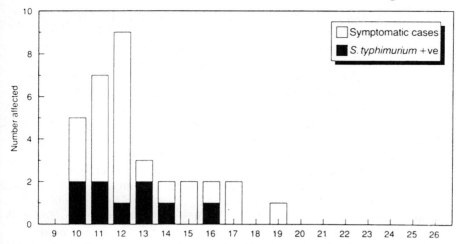

Figure 8.6 Common source outbreak of *S. typhimurium* at bank headquarters. After Breen, D. *et al.* (1988) *Communicable Diseases Scotland Weekly Report*, **22**, 6–8.

whether one is dealing with a point-source outbreak, associated with a social function, golf club outing, etc., or a common-source outbreak (Figure 8.6), associated with eating at the same restaurant or canteen, or food from the same retail outlet(s) over a period of time (which could be several days or weeks), or one that is a consequence of person-to-person spread, possibly following an unreported point-source outbreak, for example in a long-stay hospital or residential home.

A complete clinical history is always important and can often give a lead to the likely aetiological agent. When the predominant presenting symptom is vomiting, as in staphylococcal food poisoning, the incubation period is usually less than 6 hours, but may be as long as 24–36 hours in viral gastroenteritis. If diarrhoea is the presenting feature, an incubation period of 8 hours or more is likely. When the incubation period is particularly short (15 minutes–1 hour), a chemical or toxic form of food poisoning should be considered.

The microbiological investigations should be undertaken concurrently with the epidemiological enquiries. Samples of faeces from cases, suspected foods and where available swabs from kitchen utensils (knives, spoons, mixing-bowls, etc.) and work surfaces should be submitted to the laboratory as soon as possible. Specimens of vomitus are particularly valuable when staphylococcal (or chemical) food poisoning is suspected. Blood samples for serology may be helpful in the diagnosis of VTEC haemorrhagic colitis and the associated haemolytic uraemic syndrome (HUS). The appropriate laboratory should always be alerted as early as possible in the investigation of the outbreak, to enable preparations to be made to deal with the extra workload.

Assessment of the available clinical, laboratory and epidemiological findings should help direct the investigation towards a particular meal or item(s) of food.

In an explosive point-source outbreak associated with a social function, the investigation can be limited to the one (or more) meal(s) common to all those affected. Individual food 'dislikes' or dietary restrictions for personal, ethnic, religious reasons etc. should be enquired about, thus enabling certain foods to be excluded. The study of catering records or menus in restaurants, hospitals, schools, etc. may be helpful in pinpointing suspect foods. From the information obtained a food specific attack rate table can be prepared (Table 8.1).

Cross-contamination between raw and cooked foods can readily deceive the unsuspecting investigator. The location of the food preparation areas in the kitchen and the juxtaposition of cooked foods to uncooked foods (e.g. raw poultry or other meats) should be specifically enquired about, along with food-handling practices and storage (e.g. refrigeration) facilities.

Apart from point-source outbreaks, however, it may be difficult to obtain a meaningful food history. Nevertheless, the patient and family should be asked about their eating and shopping habits, what type(s) of food they eat and who their regular (and occasional) suppliers are. Any unusual items of food eaten should be noted. In every investigation it is important to keep an open mind as to the possible food vehicle. Some unusual foods which have featured in outbreaks of salmonellosis in recent years have included salami-sticks, beansprouts, chocolate, pepper and yeast flavourings. Vegetables (potatoes, etc.) and fresh apple juice have been identified in outbreaks of VTEC infection.

In the absence of confirmatory laboratory evidence, it may be necessary to undertake an analytical epidemiological study such as a cohort or case-control study, in which the food histories of both ill and well persons are compared. These techniques can be utilized to identify the most probable offending food vehicles in point-source and common-source outbreaks respectively.

8.6.1 Cohort study

Following a point-source salmonella outbreak affecting a wedding party, a food history was taken from affected guests along with a random sample of those who remained well. A food-specific attack rate table was prepared relating to individual items of food eaten or not eaten by all who were interviewed (Table 8.1). When the attack rates were calculated, the food item showing the greatest

Table 8.1 Food-specific attack rates among wedding guests

	No. of persons who ate specified food			No. who did not eat specified food		
	Total	Ill	(%)	Total	Ill	(%)
Shrimp cocktail	157	106	(68)	73	10	(14)
Turkey	210	105	(50)	20	15	(75)
Baked ham	37	16	(43)	193	94	(49)
Vegetables	158	73	(46)	76	46	(61)
Apple pie	93	39	(42)	137	80	(58)

difference in the attack rates between those who ate and those who did not eat it, was shrimp cocktail – *viz*. 68% of those who ate and 14% of those who did not eat this food. Statistical analysis showed that salmonella infection was significantly related (P=0.001) with eating shrimp cocktail; *S. java* was subsequently isolated from left-over shrimps in the hotel kitchen. The occurrence of illness affecting the 10 persons who reported not eating shrimp cocktail, may be due to faulty recollection of what they had eaten or to possible cross-contamination of other foods.

8.6.2 Case-control study

In a case-control study, it is necessary to recruit at least two controls (i.e. unaffected persons who were potentially exposed to the same risk of acquiring infection) for every one case. Controls may be nominated by the case (or parent) in the form of 'neighbourhood controls', taken from the community health index (CHI) or obtained via the GP register, but in either instance usually require to be matched with cases by age, sex and social characteristics.

Following a nationwide outbreak due to *S. typhimurium* phage type 124 (an unusual phage type in the UK), initial epidemiological enquiries suggested that a particular brand of imported salami-stick was the possible vehicle of infection. A case-control study was undertaken, the results of which showed a statistically significant association (P=0.001) between infection and eating the brand of salami-stick in the week before onset of illness. After the epidemiological results were analysed, *S. typhimurium* type 124 was isolated from a salami-stick obtained from the home of one of the cases interviewed, following which the manufacturers were advised and the product was withdrawn from sale.

Case-control studies, however, are less likely to be successful in the investigation of common-source outbreaks involving more commonly eaten foods (poultry-meat, etc.). Also it is necessary to be aware that analytical studies only identify the food(s) consumed by those affected, but these may have been contaminated from another source during preparation.

Interpretation of the epidemiological data obtained from the questioning of individual – and possibly numerous – cases in an outbreak investigation can be daunting unless undertaken properly. In recent years this has been greatly facilitated by using modern computer programmes such as '*Epi-Info*', which can compile and quickly analyse the data entered.

8.6.3 The role of reference laboratories

Epidemiological investigations of outbreaks of food poisoning in recent years have increasingly utilized the expertise provided by specialist reference laboratories.

In the UK, routine serotyping and phage-typing of isolates of *Salmonella* spp.

undertaken by the PHLS Laboratory of Enteric Pathogens, London and the Scottish Salmonella Reference Laboratory, Glasgow, have greatly facilitated the surveillance and investigation of numerous outbreaks. Other laboratory technologies such as antibiograms, biotyping, DNA probes and plasmid-profile fingerprinting have also been utilized to demonstrate a correlation between isolates of salmonella organisms from food(s), patients and/or the environment. In a large outbreak due to *S. typhimurium* phage type 49 at a psychiatric hospital in the North of England in 1984, raw poultry-meat being prepared in the kitchen was shown conclusively to have contaminated cold roast beef, which acted as the food vehicle, transmitting infection to 450 patients and staff.

Using such techniques, several community outbreaks have been identified and traced to nationally distributed foods such as imported chocolate (*S. napoli*) and salami-sticks (*S. typhimurium* phage type 124), powdered milk (*S. ealing*), beansprouts (*S. saint-paul* and *S. virchow*), yeast flavourings (*S. manchester*) and ethnically prepared poultry-meat (*S. wangata*), respectively. The observed rise in the incidence of infection by *S. enteritidis*, and in particular phage type 4, during the 1980s was similarly due to coordinated national surveillance programmes based on routine typing (Figure 8.7).

Serotyping, phage-typing, etc. and/or toxin-testing of other food-borne pathogens (*E. coli* 0157, *Campylobacter* spp., *Staph. aureus*, *B. cereus*, *Cl. botulinum*, *Cl. perfringens* and *L. monocytogenes*) in England and Wales and in Ireland are also undertaken by the Public Health Laboratory Service. In Scotland, similar facilities for the further characterization of *Campylobacter* spp. and *E. coli* 0157 are available at the Department of Medical Microbiology, the Royal Aberdeen Hospitals, Aberdeen.

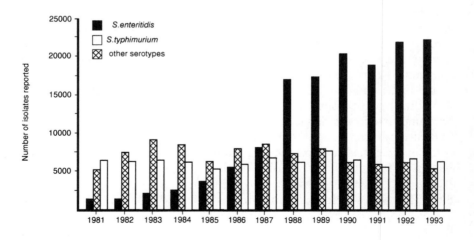

Figure 8.7 Isolates of *S. enteritidis*, *S. typhimurium* and other serotypes in the UK, 1981–1993. (Source: PHLS, London and Scottish Salmonella Reference Laboratory, Glasgow.)

8.7 SURVEILLANCE

The basis of surveillance is turning data into information, the three principal features of which are:

1. the routine collection of data;
2. its collation and analysis; and
3. the dissemination of the information to those persons who can implement the necessary control measures at a local, national or international level.

8.7.1 Surveillance → Information → Intervention → Prevention

In most countries, surveillance of food poisoning is based on statutory notifications and/or laboratory reporting of food-borne enteropathogens. In the UK, surveillance data are obtained from several complementary information sources:

1. Statutory notifications of food poisoning *per se* collated by OPCS (in England and Wales), ISD (in Scotland) and DHSS (NI) (in Northern Ireland), whether or not supported by laboratory confirmation.
2. Voluntary reporting by microbiology laboratories of isolates of food poisoning organisms from human, veterinary and environmental sources in England and Wales to the PHLS Communicable Disease Surveillance Centre (CDSC) and the Ministry of Agriculture, Fisheries and Food (MAFF), London, in Scotland to the Scottish Centre for Infection and Environmental Health (SCIEH), Glasgow and in Northern Ireland to DHSS (NI), Belfast.
3. Outbreak reporting by public health physicians and/or environmental health officers to CDSC, SCIEH and DHSS (NI) respectively.

National data and other relevant epidemiological information are published regularly in the PHLS Communicable Disease Report (CDR), the Scottish Centre for Infection and Environmental Health (SCIEH) Weekly Report and the Northern Ireland Communicable Disease Monthly Report, respectively. Supplementary forms of information feedback include area–regional newsletters to clinicians, microbiologists, infection control nurses, EHOs, etc., summarizing outbreaks and other items relating to food-borne disease of local, national and international interest.

Other countries, for example the USA (Morbidity & Mortality Weekly Report), Canada (Canada Diseases Weekly Report), Australia (Communicable Diseases Intelligence), New Zealand (Public Health New Zealand Report), France (Bulletin Épidémiologique Hebdomadaire), Denmark (Epi-Nyt), Netherlands (Staatstoezicht op de Volksgezondheid), etc., similarly publish national data at weekly or monthly intervals. Reviews of salmonella isolates and/or outbreaks of food poisoning and food-borne disease are also published periodically by some countries.

No one reporting system is however sufficient by itself and requires to be complemented by other surveillance mechanisms in order to obtain the most meaningful data (Table 8.2). Formal notifications of food poisoning are based primarily on a clinical opinion, the comprehensiveness of which may fluctuate widely among notifying practitioners. Consistency of reporting in laboratory-based systems is dependent on appropriate specimens being submitted by physicians, being successfully processed in the laboratory and thereafter reported to regional or national surveillance centres. Surveillance can be further enhanced via monitoring of foods by local health authorities (e.g. milk, water, etc.), from within the food industry (e.g. poultry industry, dairy industry, etc.), or from periodic surveys of food such as cold meats, sandwiches, etc.

Since 1989, laboratory data have been further utilized in Scotland by the development of the 'Reportable Infections' surveillance programme which embraces 32 diseases for which there is no statutory requirement to notify or otherwise report. Several of these (e.g. salmonellosis, campylobacteriosis, listeriosis, *E. coli* 0157) are commonly food-borne, and have been subjected to more active surveillance and in-depth epidemiological studies in an effort to obtain better information relating to the sources of infection and the necessary preventive measures.

In England and Wales, the Royal College of General Practitioners has developed a 'Sentinel Practice' scheme involving 60 general practices and approximately 425 000 patients, for a range of clinical conditions including infectious gastrointestinal disease. Participating practices report weekly on clinically diagnosed cases, the data relating to which are published weekly. Following the recommendations of the Richmond Committee Report in 1990–1991, a more comprehensive, but one-off, in-depth epidemiological study of Infectious Intestinal Diseases (IID) commenced in England in September 1993. The purpose of the study is to investigate the incidence, microbiological causes and costs of IID in the community, based on 72

Table 8.2 Some food-borne disease surveillance mechanisms

Clinical aspects	*Environmental aspects*
Statutory notifications of 'food poisoning'	Local health authority monitoring' of foods – milk, water, etc.
Voluntary laboratory reporting	Commercial monitoring of foods – poultry industry, dairy industry, etc.
Epidemiological investigation of outbreaks	Surveys of food preparation premises – shops, restaurants, hospitals, etc.
Hospital admissions	Surveys of food suppliers – cold meats, sandwiches, etc.
GP sentinel ('spotter') practices	Sewage effluents – human, animal (abattoir)

practices in three regions of the country. The components of the study are:

1. a population study;
2. a GP case-control study;
3. an enumeration study;
4. a study of the socioeconomic costs; and
5. a long-term follow-up of chronic sequelae of IID.

Although this study will provide valuable insights into the cause of IID, its design will not allow it to be of value in monitoring trends. It will, however, provide the most accurate estimate to date of the relationship between the true incidence of IID and that detected by routine national surveillance in England. A similar, although less comprehensive sentinel system for gastrointestinal disease has been operating in the Netherlands for several years.

The introduction of electronic communication systems in recent years such as Epinet, a UK-wide electronic mail-box system linking district health departments with each other and with the respective surveillance centres in London, Cardiff and Glasgow, has greatly enhanced and speeded-up the flow of epidemiological information. The Infectious Diseases Surveillance Systems introduced throughout Wales (IDSS) and in Scotland (SIDSS) in the past 2–3 years are examples of computerized systems for the collation of statutory notification (and other) data by public health physicians; such systems allow rapid local analysis of complex datasets, which in conjunction with Epinet have made possible the establishment of detailed, up-to-date national databases from health authority computerized data files.

In France, surveillance of acute diarrhoeal disease has been enhanced by the development of electronic reporting systems (Telematics), whereby general practitioners using home terminals are linked to regional computer bases. Advantages of the system include ease of reporting and rapid feedback to participating practitioners and public health officials.

In North America, awareness of the value of surveillance has existed for over 50 years. Information on outbreaks of food-borne disease has been collated at the US Centers for Disease Control (CDC), Atlanta since 1966, although the data are not fully comprehensive due to considerable variability in outbreak reporting from among and within the 50 states. In Canada, data on food-borne infection are more complete, with each of the 10 provinces and territories reporting annually to the Laboratory Centre for Disease Control (LCDC) in Ottawa.

As a consequence of the increased international awareness of the health and economic effects of food poisoning, and in particular in relation to tourism, the World Health Organization implemented a 'Surveillance Programme for the Control of Food-borne Infections and Intoxications in Europe' in 1980, collating national and international data on food-borne disease. Comparability of data, however, poses considerable problems due to the different surveillance systems (statutory notification, voluntary laboratory reporting, GP sentinel practices and/or epidemiological investigation of outbreaks) operating within and between different countries.

8.8 EPIDEMIOLOGY OF SPECIFIC ORGANISMS

8.8.1 Salmonellosis

Salmonellosis emerged as a public health problem in many countries including Britain shortly after the Second World War, having been introduced into the food chain via contaminated batches of dried egg imported from the USA. It was not until the mid-1960s, however, that it became an infection of increasing significance. Throughout the 1970s and 1980s, reports of non-typhoid salmonella infections and associated outbreaks continued to increase, involving a wide range of food vehicles.

An unquantifiable part of this increase can probably be attributed to improved national surveillance systems. Nevertheless, a real increase would also appear to have been taking place in many countries, with over 33 700 cases (annual infection rate of 69 per 100 000) recorded in Britain in 1993. Approximately 10–15% of salmonella infections reported in the UK each year are acquired overseas, usually involving returning holidaymakers who have been in southern Europe or further afield. Estimates of unreported cases based on studies in North America have suggested that altogether more than two million people may be infected in Britain every year.

The serotypes which have most commonly caused human infection in Britain in recent years have been *S. typhimurium, S. enteritidis* and *S. virchow*. Among other serotypes which have featured periodically have been *S. agona* (introduced in 1970 via contaminated fish-meal from Peru), *S. hadar, S. heidelberg* and *S. wangata*. In animals the more common serotypes have been *S. typhimurium* (most species), *S. dublin* (cattle) and *S. montevideo* (sheep), while among poultry flocks *S. enteritidis* and *S. virchow* have become increasingly predominant.

In 1980, less than 10% of all human salmonella infections in the UK were due to *S. enteritidis*, and by 1990 had increased to over 60% of infections, with phage type (PT) 4 strains predominating. Up until 1987, *S. enteritidis* PT 4 infections in the UK almost invariably involved tourists who had been in southern Europe. Since then there has been accumulating evidence of infection by *S. enteritidis* PT 4 affecting persons with no history of overseas travel, among whom the role of poultry-borne and subsequently also egg-borne spread, became increasingly apparent. Throughout 1988 and 1989, epidemiological and bacteriological investigations increasingly incriminated egg-based dishes, in particular those containing raw egg as an ingredient (e.g. home-made mayonnaise, mousse, ice-cream, egg-nog, etc.). Other European countries reported similar problems.

Salmonella enteritidis PT 4 was first isolated from bulk liquid egg in the UK in 1985. Since then veterinary reports and other surveillance mechanisms have increasingly demonstrated the presence of *S. enteritidis* PT 4 in broiler and laying flocks, concurrent with the upsurge in human infection. Outbreaks have been reported in the UK associated with imported eggs and poultry meat. Surveys by the PHLS in England and Wales of home-produced and imported eggs from other EC countries have demonstrated a continuing problem of salmonella contamination, with 0.9% of British produced and 1.6% of imported eggs infected; 10 000 million eggs are eaten each year in Britain. No difference was found in infection rates between battery-produced and free-range eggs. Despite official government recommendations advising against the use of raw shell eggs as an ingredient of desserts, home-made mayonnaise and other dishes not subjected to further heating, egg-related outbreaks are still being reported. Nevertheless contaminated poultry-meat continues to be the most important salmonella food vehicle.

In North America, *S. enteritidis* although of different phage types (PT 8 and PT 13a) to that prevalent in Europe, has also been increasingly identified with hen eggs and poultry-meat.

Although now less common than *S. enteritidis* in many countries, *S. typhimurium* continues to cause considerable human disease, with some strains, in particular phage types 204c and 104, both of which are primarily bovine-associated, having developed marked antibiotic resistance. *Salmonella virchow*, a particularly invasive serotype and the third most commonly recorded in humans in the UK, has frequently been isolated from poultry; multi-resistant strains have also increasingly been reported. Between mid-summer 1992 and the early months of 1993, over 350 cases of infection due to *S. wangata*, a rare serotype, were identified in the north of England and west-central Scotland affecting immigrant families from the Indian subcontinent and other persons who had eaten chicken-based dishes in ethnic food restaurants.

Case report (Sibbald et al., 1994)

On 29 July 1993, *Salmonella* group D organisms were isolated from an adult in Edinburgh who had become ill 4 days previously. The patient was visited the same day by an EHO, who noted details of food consumed in the 72 hours before the onset of symptoms.

A correlation of further *Salmonella* group D infections reported over the next few days confirmed a common restaurant venue in which nine persons had eaten lunch on the 24 July 1993, the starter dish for which was a salad of melon with mango, served with a strawberry dressing which contained raw egg yolks. A chef, who had tasted the dressing before it was served, had also suffered symptoms of salmonella food poisoning and was subsequently found to be similarly infected.

Preparation of the dressing entailed the cracking of four eggs and separating the yolks into a container into which the other ingredients were incorporated. These normally included white wine vinegar, although on this particular occasion no vinegar was used. Discussions with four chefs revealed that one of them who had prepared the dressing in question, did not use wine vinegar while the three others did. 'Quality control' took the form of several tastings as each ingredient was added. Once prepared it was left out on the work bench for convenience.

In all, 11 persons were traced who had become ill after eating at the restaurant, from nine of whom salmonella organisms, subsequently identified as *S. enteritidis* PT 4, were isolated.

8.8.2 Campylobacter enteritis

Campylobacter enteritis due to *Campylobacter jejuni/C. coli* is now the most commonly identified cause of gastroenteritis in humans. Increasingly reported throughout the 1980s, campylobacters have now overtaken salmonellas in many countries. Over 43 700 cases of campylobacter infection (mainly *C. jejuni*) were recorded in the UK in 1993 (annual infection rate of 78 per 100 000), with rates exceeding 150 per 100 000 in some areas. To what extent the four-fold rise in the reported incidence of campylobacter infections between 1980 and 1993 is due to enhanced awareness, surveillance programmes and/or improved reporting, is unclear. Elsewhere in Europe and North America an increasing number of countries are now reporting campylobacter infections. In New Zealand, national rates of 150 per 100 000 have regularly been recorded in recent years, with rates exceeding 300 per 100 000 in Canterbury province in the South Island.

In developed countries, campylobacter infections affect all age groups, with peaks of increased incidence occurring among pre-school children and young adults.

The seasonal distribution shows a characteristically distinct peak during late spring and early summer, with a secondary late autumn rise (Figure 8.1). Similar observations have also been reported from other countries. The reasons are not clear, but may be due to an inter-play of environmental, ecological, agricultural and/or food marketing factors.

The epidemiology is also much less clear than with the salmonellas, and despite increasing evidence of food-borne transmission, campylobacters continue to be under-represented in most surveillance programmes. This is, in part, due to the lack of readily available laboratory typing facilities, but also to the relatively shorter time (hours rather than days) that *Campylobacter* spp. are capable of surviving on foods. In addition most infections present sporadically as single cases or as household outbreaks, many of which are not routinely investigated. In consequence food vehicles are less frequently identified, although outbreaks involving undercooked chicken, rare meats, raw milk and unchlorinated water supplies, have been reported from the UK, USA, the Netherlands and Australia, among others. Pig meat is an important source of *C. coli* infection in Europe, particularly where eating smoked or undercooked pork is popular.

There is some evidence to suggest that much of the continuing rise in the number of reported cases in the UK is due to the increasing consumption of fresh poultry-meat, a popular and relatively cheap source of protein. The handling of raw chickens during food preparation has been associated with acquiring campylobacter infection, while cross-contamination within kitchens has also been demonstrated. Case-control studies in the USA and in the Netherlands have demonstrated that up to 50% of infections are associated with the consumption of chicken. In one outbreak in Australia, involving diners who had eaten chicken casserole, *C. jejuni* organisms of the same serotype were isolated from patients and from samples of undercooked fresh chicken obtained at the restaurant.

While surviving readily within the refrigerator, campylobacters present on foods are readily destroyed by conventional cooking, but may survive in the centre of made-up, comminuted meats such as meat-patties. Salads, cooked meats and other ready-to-eat foods have all been implicated in household infections, but were probably cross-contaminated via unwashed hands, utensils, etc. from raw foods present within the kitchen.

Cattle, sheep, pigs and other domestic animals, including household pets, frequently harbour campylobacter organisms in their intestinal tract. Human infections, particularly in rural areas, are frequently acquired by drinking raw milk and/or by direct zoonotic spread involving farming families and abattoir workers. In early 1991 in north-east Scotland following an educational farm visit, which included handling young lambs and calves and being given untreated milk to drink, a group of school-children became ill over the next 2–3 days; *C. jejuni* was isolated from eight children,*Cryptosporidium* spp. from three and both organisms from four children respectively (J. Curnow, personal communication). Contamination of carcasses during processing after slaughter has frequently been demonstrated. *Campylobacter* spp. have been isolated from pork, beef and veal carcasses in Canada, from samples of beef, pork, lamb and from 30–47% of edible offals (liver, heart and kidney) in surveys in England. Water-borne outbreaks associated with untreated or inadequately treated drinking water supplies, have been reported from several countries including the UK, USA and Sweden.

There has been increasing epidemiological and laboratory evidence from several urban areas in the UK (including Edinburgh, Plymouth, Gateshead, Wakefield and South Wales) suggesting an association, during the months of May and June, between campylobacter illness and the consumption of milk from doorstep bottle deliveries, the tops of which had been attacked by magpies or jackdaws.

Case report (Brown et al., 1988)

Following a business lunch at a hotel attended by 51 delegates from all over the north of England, 24 persons (47%) became ill 1–5 days later. Faecal samples were obtained from 17 of those affected, from seven of whom *Campylobacter* spp. was isolated. Serotyping showed that several different serotypes were involved, a not uncommon finding in the investigation of campylobacter outbreaks.

A food history analysis showed that although several of the meat dishes served had higher attack rates for those persons eating those dishes, no one food was found to be significantly associated with illness. The hotel kitchen was modern, had recently been extended, and was well organized and managed by the catering staff. Refrigeration facilities were good, segregation of raw foods and cooked foods was generally satisfactory and methods of food preparation appeared good. Subsequent on-going observations of routine food-handling practices in the kitchen showed that while the chef usually washed his hands between raw and cooked foods, on one occasion while under considerable pressure he failed to do so between handling raw chickens and cooked vol-au-vents, a lapse which could readily result in cross-contamination of a sufficient number of campylobacter organisms to cause an outbreak.

8.8.3 Listeriosis

Outbreaks of listeriosis have been reported in the former Czechoslovakia, Germany, New Zealand, Australia, France and the USA since the mid-1950s, although their origins and routes of transmission were seldom ascertained.

During the 1980s, several large outbreaks with fatality rates of up to 27%, were reported in Canada, the USA and Switzerland, due to contaminated coleslaw, pasteurized milk and soft cheeses respectively (Table 8.3). An extensive nationwide outbreak of listeriosis in Britain during 1987–1989 was associated with imported pâté, while in France in each of 1992 and 1993, two distinct outbreaks occurred due to contaminated cold pork products. Sporadic cases of infection have also been associated with eating uncooked hot-dogs and undercooked chicken, soft cheeses and cook-chilled food.

Despite the evidence of contamination of a wide range of foods in various surveys, a causal relationship with human illness in Britain has, however, only occasionally been established in incidents in which cheese (imported soft cheese and UK goats' milk cheese), cook-chilled chicken and vegetable rennet were identified. Concern about 'cook-chill' meals in particular is understandable in view of the rapid expansion in the use of these and other pre-cooked foods, and in their consumption by vulnerable groups of the population (pregnant women, the elderly and the immunocompromised), both in the home and in hospital.

Table 8.3 Listeriosis: reported food-borne outbreaks

	Country	Food vehicle	No. affected
1981	Canada	Coleslaw	41
1983	USA, Massachusetts	Pasteurized milk	49
1983–1987	Switzerland	Cheese	122
1985	USA, California	'Mexican-style' cheese	142
1987–1989	UK	Pâté	355
1992	France	Pork tongue in jelly	279
1993	France	Pork rillettes	25

The serotype strains of *L. monocytogenes* responsible for the 1988–1989 'epidemic' in Britain were demonstrated in samples of imported pâté, and in particular from one manufacturing plant. Government health warnings issued in the summer of 1989 drew attention to the dangers to vulnerable population groups of eating certain high-risk foods (soft cheeses, cook-chill foods, pâté).

Product withdrawal of the incriminated pâté and the more stringent production requirements placed on the manufacturers, resulted in the incidence of listeriosis thereafter returning to the lower 'pre-epidemic' levels of the earlier 1980s.

Case report (Bannister, 1987)

A 36-year-old woman was admitted to hospital in London with an 18-hour history of fever, back pain, aching legs and increasing headache with neck stiffness. She had previously been in good health without any debilitating condition, altered immunity or neurological disease. She was not pregnant nor taking any medication. Blood cultures (×3) and microscopic examination of the cerebrospinal fluid (CSF) were negative; culture of the CSF, however, succeeded in growing small rod-shaped bacteria subsequently confirmed as *L. monocytogenes*. Her condition improved within a few hours of antibiotic treatment being changed to chloramphenicol.

On further recovery it was elicited that she had purchased two soft French cheeses for Christmas, and had kept them refrigerated for 2 weeks before eating some of each during the week before her illness. Neither her husband nor son, both of whom remained well, had eaten any. Remaining portions of the cheeses were examined at the PHLS Food Hygiene Laboratory, Colindale, where a heavy growth of *L. monocytogenes* was produced from a soft country cheese; the other cheese, a Camembert, did not yield any growth of significance. Examination of unopened samples of the same soft country cheese failed to grow *L. monocytogenes*.

The strain isolated from the patient's CSF and that from the cheese were both shown at the reference laboratory to be of serotype 4, and subsequently to be of indistinguishable phage types.

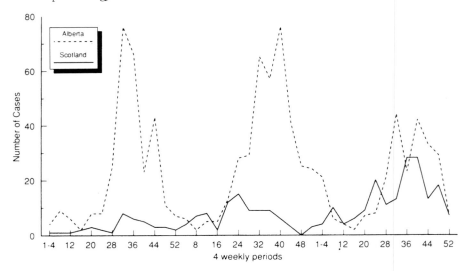

Figure 8.8 Seasonal distribution of *E. coli* 0157 reported in Alberta and Scotland, 1988–1990. (Source: Alberta Health, Edmonton and CDS Unit, Glasgow.)

8.8.4 Haemorrhagic colitis (VTEC)

Verotoxin-producing strains of *E. coli* (VTEC), in particular serotype 0157:H7, were first recognized in North America in the early 1980s as a cause of haemorrhagic colitis and the associated haemolytic uraemic syndrome (HUS). Community outbreaks of bloody diarrhoea in two different states in the USA in 1982 were associated with eating undercooked hamburgers from a fast-food restaurant chain. Coincidentally in Canada, a study of sporadic cases of childhood haemolytic uraemic syndrome (HUS) demonstrated an association with faecal cytotoxin and VTEC in the stools of those affected.

Throughout the 1980s in Canada, where the disease is notifiable nationally, an exponential rise occurred in the number of isolations of *E. coli* 0157 submitted to the National Reference Centre for Enteric Bacteriology at LCDC in Ottawa. The number of infections doubled each year between 1982 (25 isolates) and 1989 (2407), although it has decreased again in more recent years (H. Lior, personal communication). Within Canada the highest infection rates are reported from Alberta. In the USA where national surveillance of *E. coli* 0157 infection was introduced in 1993, the trends to date have been less apparent.

Most infections occur during the summer months (the barbecue season) when a marked increase is seen in the number of reported cases. Cases of HUS show a similar seasonal distribution. The majority of infections occur sporadically or within family groups, without any food source being ascertained. Community and point-source outbreaks have been reported, including several which were food-borne (Table 8.4).

Table 8.4 *E. coli* 0157: reported food-borne outbreaks

	Country	Food vehicle	No. affected (HUS)
1982	USA	Ground beef*	47
1984	USA	Ground beef*	34 (1)
1985	UK	Raw potatoes	24
1986	Canada	Raw milk	43 (3)
1986	USA	Ground beef	37 (4)
1987	UK	Turkey-roll	26 (1)
1987	Canada	Ground beef*	15
1988	USA	Meat-patties	54
1990	UK	Various foods	16 (4)
1991	UK	Burgers	23 (3)
1991	UK	Yoghurt	16 (5)
1992–1993	USA	Burgers*	583 (41)
1993	UK	Raw milk*	6 (3)
1993	UK	Beefburger*	8 (1)
1994	UK	Pasteurized milk*	100+ (9)

* Laboratory confirmed. HUS, haemolytic uraemic syndrome.

Among these were several clinically severe nursing home outbreaks, in one of which in Alberta in 1987, examination of raw frozen ground beef (hamburger patties) from the same consignment yielded the identical strain of *E. coli* 0157 as had been isolated from affected residents. In the largest recorded outbreak worldwide, over 500 persons in western USA were affected following eating hamburgers at a fast-food restaurant chain. Surveys in North America have demonstrated levels of contamination of various raw meats (beef, pork, lamb and poultry) sampled at meat-processing plants, restaurants and retail outlets such as butchers' shops.

Outbreaks have also been reported in association with untreated milk, along with water-borne transmission and person-to-person spread.

In the UK an increasing number of cases and outbreaks have been reported since 1989 in parallel with more laboratories examining for *E. coli* 0157. The highest infection rates have been recorded in Scotland with a peak of 202 cases (4 per 100 000) in 1991. Only in Canada have higher rates been reported. Several food-borne outbreaks affecting hospitals and other health care units (nursing homes, etc.), restaurants, retail outlets (butchers' shops, delicatessens, etc.) have been recorded. Consumers of dairy produce (milk and yoghurt) have also been affected in community outbreaks in recent years (Table 8.4). In a restaurant outbreak near Edinburgh in 1990, 16 persons were affected, including four children who developed HUS; despite detailed food histories no specific vehicle was identified, with cross-contamination from a raw food within the kitchen the most likely contributory factor. In 1991, 24 cases of haemorrhagic colitis mainly in north-west England, due to an unusual strain of *E. coli* 0157 were associated with eating 'burgers' from a fast-food restaurant chain; in-depth microbiological investigation of the raw meat supply, however, failed to isolate the organism. Since then, isolates of *E. coli* 0157 have success-fully been made on several occasions from food in the UK (viz. from milk and raw hamburger meat) following the investigation of outbreaks. Abattoir surveys have shown a correlation between infected cattle and subsequent contamina-tion of carcasses, while veterinary studies demonstrating a zoonotic link between cattle and human infection have further confirmed the bovine reservoir of VTEC infection.

Outside North America and the UK, an increasing number of other countries are also now identifying and reporting VTEC infection along with outbreaks of HUS, in many instances involving verotoxin-producing *E. coli* serotypes other than 0157.

Case report (Davis et al., 1993)

Between mid-November 1992 and early February 1993, over 500 laboratory-confirmed cases of *E. coli* 0157:H7, mainly children, were reported across four different states in western USA. Over 170 persons were admitted to hospital, 41 of whom developed HUS including four children who died. Case-control studies undertaken in Washington, Nevada and California implicated eating regular-sized hamburgers at chain A restaurants during the week before onset of symptoms. Routine investigations in Idaho revealed that 13 of the 14 cases identified in that state had also eaten at a chain A restaurant. On January 18th, a multi-state recall of unused hamburger patties from chain A restaurants was initiated, with the number of new cases occurring decreasing quickly thereafter.

Investigations at chain A restaurants resulted in the isolation of *E. coli* 0157:H7 from several batches of hamburger patties produced on consecutive dates in November, and distributed to restaurants in affected states. Further characterization of the isolates showed that these were identical to the outbreak strain causing human illness.

Extensive investigations along the food chain failed to reveal the source of infection of the contaminated patties. Slaughtering and carcass preparation procedures appeared to have been the most likely means by which meat was contaminated, with undercooking of hamburgers in chain A restaurants result-ing in the outbreak taking place.

8.8.5 Viral infections

Viral food-borne infections are mainly of human origin, either as the result of direct spread from an infected food-worker or indirectly via shellfish farmed in sewage-polluted waters. Small round structured viruses (SRVSs) such as the Norwalk agent, and small round viruses (SRVs) are the most frequently identified causes of viral gastro-enteritis. It is only within the past 10–15 years, however, that their role in food poisoning has been recognized. The increase in laboratory reports of viral gastroenteritis during the 1980s has been in part due to more frequent examination of specimens by electron microscopy in the investigation of outbreaks, but more importantly to an enhanced awareness of the need to consider a viral aetiology and to submit faeces to the laboratory within 24–48 hours of the onset of symptoms.

Molluscan shellfish (cockles, mussels, oysters) have been among the most frequently identified food vehicles, in particular where farmed in sewage-polluted waters. Oysters are usually eaten raw and are especially hazardous. A 40-year review of shellfish-associated disease in England and Wales showed a marked increase in outbreaks with a viral or suspected viral aetiology during the 1980s, the majority of which were caused by eating inadequately cooked cockles or raw oysters.

Outbreaks of viral food poisoning due to contaminated foods other than shellfish, such as salads and cold foods, have also been reported. In a hotel outbreak of gastroenteritis due to SRSVs, it appeared that projectile vomiting from a food-handler was responsible for contaminating the kitchen environment, including work surfaces and various foods.

Case report (Arnold et al., 1989)

Several cases of acute gastroenteritis presented on a Monday morning to the primary care centre of a military establishment in Scotland. All of those affected were officers who had attended a social function held on the Friday evening, at which two buffets each with a range of over 90 foods were provided along with an 'oyster bar' serving champagne, raw oysters and crab claws.

All together 37 persons were identified who had experienced diarrhoea (31 persons), vomiting (27), abdominal pain (25) and/or nausea (20), developing between 12 and 67 hours (mean 38 hours) after the function with symptoms lasting between 6 and 36 hours; fever (11) and headache (8) were less commonly reported. A wide variety of foods from the buffet had been consumed, although it quickly became apparent that illness was most commonly associated with eating oysters from the 'oyster bar'. This was further supported by a simple food-associated attack rate study, with no other foods consumed showing any strong association with illness.

The oysters had been imported frozen from Japan in 1-kg bags with no English labelling other than the name of the importer. Translation of the Japanese instructions subsequently read '...when defrosted fry or sauté with butter, cook at high temperature' – advice regarding which the catering manager was unaware.

Microbiological examination, including virology, of left-over seafood samples was inconclusive, as were the only two stool samples submitted. Despite a lack of supportive laboratory evidence, the epidemiological findings were sufficiently strong to implicate uncooked oysters as the food vehicle of the outbreak, while the clinical picture indicated a viral aetiology, possibly SRSVs.

9 Microbiological control of food production

T.A. Roberts

Throughout the ages man has tried to control food-borne diseases and spoilage by, for example, learning not to eat some foods that had putrefied because they were sometimes toxic, and by evolving ways of preserving foods in an edible condition. Empirical knowledge of food preservation processes is centuries old, and most processes used today to preserve foods were developed as a result of experience. For example, in ancient Egypt, cattle were domesticated predominantly to provide milk, which was converted into cheese following a schedule similar to a modern code of practice. Pork was cured, fish were cooked and eaten or stored salted or dried. Some 2000 years BC several different types of breads were baked, beer was brewed and several different wines made. Plants such as lettuce, cucumber, beans and cabbage were cultivated, while onions, leek and garlic were popular. Preserved foods were stored from times of plenty to times of scarcity. It is only within the last 200 years, however, that we have come to understand that many of the processes which cause deterioration of plant and animal tissues involve micro-organisms.

Although micro-organisms were observed and described by Leeuwenhoek in 1683, not until 1837 did Pasteur first associate bacteria with food spoilage. The demonstration that diseases are transmitted via foods also came in the 19th century. Hence, for most of man's history, food spoilage and disease transmission have been dealt with in ignorance of the responsible agents.

9.1 PRIMARY CONTAMINATION

When agricultural products are harvested or animals slaughtered, the numbers and types of micro-organisms they carry, and which comprise the primary contamination, vary from one commodity to another. This variation depends on the geographical region, on production methods, and on harvesting and slaughtering methods. Much effort has been put into characterizing this initial microbial flora, in the expectation that it would be possible to exclude undesirable micro-organisms, particularly those that are later associated with spoilage of the product or with food poisoning that can be traced to that product. The conversion of plant and animal tissues into food must prevent natural contaminants from either decomposing the produce or rendering it dangerous to eat because of the growth of micro-organisms able to cause illness either directly, or via toxins produced during multiplication.

The micro-organisms that become important in food poisoning or the spoilage of stored agricultural products are not necessarily relevant to the growth of animal or plant. Conversely, some organisms which do affect growth may be of little significance to the food production process: many plant diseases cannot be transmitted to humans and are, therefore, of consequence only to the grower. Some animal diseases, though, are transmissible to humans via food products and in some areas preventive action by producers has had a significant effect on human health. Eradication of animal diseases such as brucellosis and bovine tuberculosis has revolutionized the pattern of human diseases traceable to meats in some countries. It is worth noting, however, that the factor most likely to persuade animal producers to take preventive action is that sick animals may die or not gain weight as fast as healthy animals. Care with particular procedures may limit, or even reduce, primary contamination, but eventually a point is reached where the costs of those preventive measures outweigh the benefits.

Even healthy animals, however, may still carry low numbers of bacteria able to cause illness in humans. Current methods of transporting and slaughtering animals are failing to minimize the dispersal of those pathogens into the food chain. From the point of view of the food processor, the primary microbial contamination of animals and plants is rarely under such control that there is any real assurance of complete freedom from particular hazardous micro-organisms. In some instances, intensive production has markedly worsened the microbiological conditions of the raw product. A small slaughterhouse can be entirely adequate for small numbers of animals but totally unsuited to large numbers. Nor does mechanization necessarily improve hygiene or the microbiological condition of foods. Modern mechanized slaughterhouses have not provided raw meat that is less heavily contaminated with bacteria, partly because the emphasis remains on high throughput rather than on high standards of hygiene. Similarly, crops raised for local consumption present minimal spoilage problems, but if intended for a distant market, special care is often needed to protect them from further contamination and damage during transport.

Efforts continue to minimize the occurrence of primary contamination with micro-organisms able to cause illness, such as salmonellae, but attempting to control this initial flora is often less cost-effective than applying well-established preservative measures. Given the limitations of available technology, there seems little prospect that it will ever become possible to ensure that food materials are completely free of contamination by particular organisms before processing; indeed, intensive animal production has in some cases worsened the situation considerably, e.g. with respect to salmonellae.

9.2 PARAMETERS OF CONTROL

The microbial flora developing on a food is a function of the initial flora, any process that has been applied, particular properties of the food (e.g. substrates available for microbial growth) and the conditions under which it is stored. Micro-organisms need suitable combinations of water, nutrients, appropriate temperatures and pH values to multiply. The inherent properties of the food with respect to pH, water activity (a_w), the substrates available for microbial growth, and temperature largely determine which micro-organisms among those present initially can multiply and constitute the 'spoilage flora'. The safety and shelf-life of foods can be assured by manipulating those and other factors, e.g. heating for sufficient time to kill micro-organisms of concern, removing water by drying or making it unavailable for microbial growth by the addition of solutes, reducing the storage temperature, or reducing the pH either by the direct addition of acid or by fermentation.

During processing micro-organisms are killed, inhibited, removed or otherwise excluded. Process failure may lead to survival of potentially dangerous micro-organisms or their toxins, while time/temperature abuse can result in multiplication of pathogenic bacteria and moulds. The most commonly used food processes developed empirically are chilling, freezing, pasteurizing, canning, drying, salting, sugaring, acidification, fermenting, and the use of chemical preservatives such as curing salts. Nevertheless, the development of novel food processes has been greatly aided by research into the relevant properties of food-borne micro-organisms, such as their response to heat and cold, water requirement, their sensitivity to acids and alkalis and antimicrobial substances.

Organisms differ greatly in their temperature requirements (Table 9.1), and in their ability to withstand heat and cold. At a given lethal temperature, the time taken to reduce the viable numbers of micro-organisms 10-fold is termed the decimal reduction value (*D* value). Representative values are given in Table 9.2. Inactivation occurs more rapidly with increasing temperature. The death

Table 9.1 Cardinal temperatures for prokaryotic micro-organisms

Group	*Growth temperature (°C)*		
	Minimum	*Optimum*	*Maximum*
Thermophiles	40–45	55–75	60–90
Mesophiles	5–15	30–45	35–47
Psychrophiles	−5–+5	12–15	15–20
Psychrotrophs	−5–+5	25–30	30–35

(Source: ICMSF (1980a) – see recommended reading.)

Table 9.2   Heat resistance of bacteria and bacterial spores

Micro-organism	Temperature (°C)	D value (min.)
Salmonella spp.	65.5	0.02–0.25
Staphylococcus aureus	65.5	0.2–2.0
Yeasts, moulds and spoilage bacteria	65.5	0.5–3.0
Spores of mesophilic aerobes		
Bacillus cereus	100	5.0
B. subtilis	100	11.0
Spores of mesophilic anaerobes		
Clostridium perfringens	100	0.3–20.0
Cl. botulinum	100	
type A and type B proteolytic strains	100	50.0
types E and non-proteolytic types B and F strains	80	ca. 1.0*
Spores of thermophilic aerobes		
Bacillus stearothermophilus	120	4.0–5.0

* Recovery of heated spores in the presence of lysozyme increases the apparent heat resistance and the D value.
(Source: ICMSF (1980a) – see recommended reading.)

rate of vegetative bacteria increases 10-fold for every (approximately) 5°C increase in temperature within the lethal range, and that of spores for every 10°C increase. Bacteria that prefer low temperatures (psychrophiles) are unable to withstand even the modest temperatures (40°C) at which mesophiles multiply. Mesophilic vegetative bacteria, including all the common food-borne pathogens, are inactivated rapidly by temperatures above approximately 70°C, unless heating occurs at low water activities, e.g. as occurs in chocolate production, when the heat resistance increases very substantially.

Bacterial spores (endospores) occur naturally in the environment, and are consequently found in all agricultural products and on raw meats. The most certain method of controlling spores in foods is to inactivate them by heating. Their resistance to heat varies considerably with the species and is affected by the environment in which they are heated. For example, products of low pH can be rendered microbiologically stable and safe by lower heat processes than those of neutral pH, because many bacteria and spores are more sensitive to heat under acid conditions, while others are unable to grow at very low pH values (e.g. *Clostridium botulinum* will not grow below pH 4.5 (approximately) except under unusual circumstances). Hence pH plays an important role in food preservation, and numerous acids are used in food products to render them more stable and safer. Fermented foods, which include dairy, meat and vegetable products, have an excellent safety record in all parts of the world. They are generally only involved in food poisoning when the initial fermentation has been slow, resulting in multiplication of pathogenic bacteria before the inhibitory pH values are attained.

Table 9.3   The limits of pH allowing growth of various micro-organisms in laboratory media adjusted with strong acid or alkali

Micro-organism	Minimum pH	Maximum pH
Gram-negative bacteria		
Escherichia coli	4.4	9.0
Salmonella paratyphi	4.5	7.8
Vibrio parahaemolyticus	4.8	11.0
Gram-positive bacteria		
Bacillus cereus	4.9	9.3
Clostridium botulinum	4.7	8.5
Enterococcus spp.	4.8	10.6
Staphylococcus aureus	4.0	9.8
Yeasts		
Candida pseudotropicalis	2.3	8.8
Saccharomyces spp.	2.1–2.4	8.6–9.0
Moulds		
Aspergillus oryzae	1.6	9.3
Penicillium variabile	1.6	11.1
Fusarium oxysporum	1.8	11.1

(Source: ICMSF 1980a – see recommended reading.)

Table 9.4   Approximate minimum levels of water activity (a_W) permitting growth of micro-organisms at near-optimal temperatures

Micro-organism	a_W
Bacteria*	
Bacillus cereus	0.95
B. subtilis	0.90
Clostridium botulinum type A	0.95
Cl. botulinum type B	0.94
Cl. botulinum type E	0.97
Cl. perfringens	0.95
Escherichia coli	0.95
Salmonella spp.	0.95
Staphylococcus aureus	0.86
Vibrio parahaemolyticus	0.94
Yeasts	
Zygosaccharomyces bailii	0.80
Saccharomyces rouxii	0.62
Moulds	
Aspergillus candidus	0.75
Penicillium chrysogenum	0.79
Rhizopus nigricans	0.93

*a_W adjusted with salts.
(Source: ICMSF 1980a – see recommended reading, and Troller, J.A. and Christian, J.H.B. (1978) *Water Activity and Food*, Academic Press, London.)

Table 9.5   Minimum growth temperatures for food poisoning bacteria

Organism	Growth temperature (°C)
Enteropathogenic *Escherichia coli*	10
Clostridium botulinum (proteolytic)	10
Cl. perfringens	10
Bacillus cereus	8–10*
Staphylococcus aureus (toxin production)	8–10
Staph. aureus (growth)	6.7
Salmonella spp.	6.7
Cl. botulinum (non-proteolytic)	3.3
Vibrio parahaemolyticus	2–5
Streptococcus spp.	1
Listeria monocytogenes	−0.5
Yersinia enterocolitica	−2

*Psychrotrophic strains growing below 5°C have been reported.

Similarly, the storage temperature is an important factor in selecting which micro-organisms grow on a stored food. With a fuller understanding of the factors determining spoilage, it can be seen that the effort put into defining the initial flora was, in part, misplaced, because the storage conditions select those organisms able to grow, and those able to grow fastest dominate. Many of the

Table 9.6   Effects of handling and processing on micro-organisms

Operation	Food	Intended effect
Cleaning, washing	All raw foods	Reduces numbers of micro-organisms
Antimicrobial dipping/ washing	Mostly fruits, vegetables	Kills selected micro-organisms
Chilling (below 10°C)	All foods	Prevents growth of most pathogenic bacteria; slows growth of spoilage micro-organisms
Freezing (below −10°C)	All foods	Prevents growth of all micro-organisms
Pasteurizing (60–80°C)	Milk, wines, etc	Kills most non-sporing bacteria, yeast and moulds
Blanching (95–110°C)	Vegetables, shrimps	Kills vegetative bacteria, yeast and moulds
Canning (above 100°C)	Canned foods	Commercially sterilizes food; kills all pathogenic bacteria
Drying	Fruit, vegetables, meat, fish	Halts growth of all micro-organisms when a_w <0.60
Salting	Vegetables, meat, fish	Halts growth of many micro-organisms at ca. 10% salt
Syruping (sugars)	Fruits, jam, jellies	Halts growth when a_w <0.70
Acidifying	Fermented dairy and vegetable products	Halts growth of most bacteria (effects depend on acid type)
Irradiating	Various	Inactivates vegetative bacteria or spores according to dose

(Source: ICMSF (1988) – see recommended reading.)

bacteria present on red meat carcasses are mesophiles, but many perishable products derived from the carcasses are stored chilled, so the mesophiles are unable to grow. Consequently micro-organisms present as a small fraction of the initial flora multiply and become numerically dominant during storage. Microbial response to pH, a_w, and storage temperature, considered individually, is illustrated in Tables 9.3–9.5. Table 9.6 provides an overview of the impact on micro-organisms of the various individual techniques of food handling and processing.

One point that should be emphasized is that temperature plays a critical role at every stage of food processing. As traditional products and processes are changed to accommodate the demand for improved textural properties after lower heat processes, for curing with reduced levels of salt and nitrite, and for foods from which established preservatives have been removed, greater reliance is increasingly being placed on temperature control as the main factor in controlling the growth of spoilage organisms, and those able to cause food poisoning (section 10.2.2). These demands will increase as distributors, retailers and consumers seek foods with minimal additives and preservatives, yet with a greatly extended shelf-life. This is the challenge that faces the food industry in the immediate future, and one which consumers will expect it to meet at little or no increased cost.

9.3 APPROACHES TO THE CONTROL OF FOOD SAFETY AND QUALITY

Ordinances, codes of practice and laws concerning the processing, handling and sale of foods have been developed by local, national and international bodies to protect the public from adulteration, fraud and illness. Most are largely based on inspectional procedures, supplemented by microbiological testing, despite the fact that it is widely recognized that both these approaches have serious shortcomings. The food processor generally attempts to comply with relevant laws via in-house quality control departments, with personnel observing operations and making physical, chemical and microbiological tests. Some activities of quality control departments will bear little or no relationship to microbiological hazards. Recently, a more efficient approach to controlling microbiological hazards has been developed, based on the recognition of hazards associated with the production or use of foods, and the identification of points where the hazards can be controlled.

9.3.1 The inspectional approach

To judge whether food processing operations meet commercial requirements and comply with the law, quality control personnel and law enforcement officials have developed complex systems for inspecting those operations for compliance with accepted good practices. The laws the inspector must enforce, and with which the owner of a food operation must comply, often contain vague terms that do not specify exactly what constitutes compliance. Phrases such as 'cleaned as frequently as necessary', 'appropriate containers', 'where necessary to prevent introduction of undesirable micro-organisms', 'adequate methods for cleaning and sanitizing', are open to many different interpretations. Such lack of specificity and absence of indication of the relative importance of requirements leaves much to interpretation and to the discretion of the inspector. It is not uncommon for the regulation and/or the inspector to fail to distinguish between important and relatively unimportant requirements. Factors critical to safety may be overlooked or underestimated, with no differentiation between factors critical to food safety and those that are merely aesthetic.

Visits by inspectors are necessarily occasional, so their observations relate almost exclusively to what is occurring on the day of the visit at the time the operation is observed. Examination of process or laboratory records is increasingly becoming part of the inspection, but what the inspector chooses to review or emphasize is often a matter of opinion, and may not be of real importance in controlling microbiological hazards. In some countries, inspection of facilities and operations is continuous (e.g. slaughtering and meat processing). While this obviates some of the above criticisms, even full-time inspectors cannot inspect every relevant part of the process in a complex food processing operation. In control of slaughtering, the emphasis tends to be on animal health via ante- and post-mortem inspection of animals and carcasses, because the primary concern has historically been to keep meat from diseased animals out of food channels, and to ensure that slaughter and processing operations are performed under sanitary conditions. Few activities are connected directly with the control of the microbiological hazards associated with today's illnesses. Despite continuous inspection, a significant proportion of food poisoning incidents can be traced back to red meats and poultry, because the inspection cannot detect carriers of salmonella, campylobacter, etc. and current slaughterhouse practices do not prevent their contaminating the carcass. Subsequent mishandling of raw meat in kitchens of food service establishments and homes may then lead to food poisoning.

Despite these obvious shortcomings, the inspectional approach to the control of microbiological hazards dominates throughout the world.

9.3.2 Microbiological testing

Microbiological testing as a means of assessing whether a product is microbiologically hazardous is relatively recent. Microbiological criteria have been applied successfully to drinking water to protect public health, but there are few examples of such success in food control.

Testing to control microbiological hazards in foods has serious limitations. It is difficult to take samples representative of the production batch and to examine a sufficient number of sample units to obtain meaningful information about the microbiological status of a batch of food. Available microbiological methods are relatively slow, are frequently not sufficiently specific, and are expensive. It is not practicable to hold perishable products while waiting for the result of microbiological analyses. More seriously, microbiological testing only identifies the consequences and does not identify or control the causes. It is not unknown for the industry to become aware of microbiological problems in its product only after spoilage in the market-place or reports of illness. Where microbiological quality control programmes have existed, there has been heavy emphasis on testing the finished product, which has proven to be an inefficient approach to control. The unsatisfactory nature of casual sampling, the statistical basis of sampling plans and their limitations are elaborated in ICMSF, 1986 covering a wide range of foods. Sampling plans pertinent to a more recent hazard, *Listeria monocytogenes,* are described in ICMSF, 1994a – see recommended reading.

9.3.3 The hazard analysis critical control point (HACCP) approach

The HACCP concept is a systematic approach to hazard identification, assessment and control. It was first presented in outline form at the 1971 National Conference on Food Protection. A hazard analysis evaluates all procedures in the production, distribution and use of raw materials and food products, in order to:

1. identify potentially hazardous raw materials and foods that may contain poisonous substances, pathogens or large numbers of food spoilage microorganisms, and/or that can support microbial growth;
2. identify the potential sources and points of contamination by analysing each step in the food chain;
3. determine the potential for microbial survival or multiplication during production, processing, distribution, storage and preparation for consumption; and
4. assess the risk and the severity of hazards identified.

HACCP avoids many weaknesses inherent in the inspectional approach and the shortcomings of reliance on microbiological testing. It focuses attention on factors that directly affect the microbiological safety and quality of a food, eliminating wasteful use of resources on superfluous considerations. By concentrating on the control of key factors that affect safety and quality throughout the food chain, government inspectors, the producer, the processor and the ultimate user of food has greater assurance that intended levels of safety and quality are met and maintained. If a food is produced, processed or used in accordance with the HACCP system, there is greater assurance of its microbiological safety and quality. Regulatory efforts can then be expended at other facilities or operations where proper control is not exercised. The system is applicable to all parts of the food chain, from production through processing, transporting and retailing to ultimate use in the food service establishment or the home.

The HACCP system comprises six sequential steps, which are outlined below.

(a) Identification of hazards and assessment of their severity.

This includes risks associated with growing, harvesting, processing/manufacturing, distribution, retailing, preparation and/or use of raw material or food product. 'Hazard' may mean unacceptable contamination, growth and/or survival of micro-organisms which are of concern as regards safety or spoilage, and/or the unacceptable production or persistence in foods of products of microbial metabolism, e.g. toxins, enzymes and biogenic amines.

Epidemiological information gives the best evidence that a hazard exists in a given product. If food poisoning is traced to consumption of a particular product, there is no doubt that an uncontrolled hazard exists. Its source must then be determined and remedial measures implemented. For example, precooked roast beef was a common vehicle of human salmonellosis in the USA. Initially the problem was traced to inadequate heating. Safe time/temperature parameters for destruction of salmonellae were established, but despite regulations requiring the heat process, outbreaks continued. Cross-contamination between raw and cooked products was identified as a main cause of these outbreaks, with slow cooling and poor control of storage temperatures also contributory factors. With the establishment of time and temperature controls for cooking, the prevention of cross-contamination between raw and cooked products, and rapid cooling, further outbreaks were prevented.

The analysis of hazards must be quantitative to be meaningful. This requires the assessment of risk and severity. Risk relates to the likelihood of the occurrence of a hazard. Will it always be present? Will it occur once a day, or once a year? Severity relates to the magnitude of a hazard. Is it life-threatening? Will many persons become ill? Will it certainly result in extensive and serious product spoilage? Alternatively, are the consequences trivial or insignificant in terms of food poisoning or product spoilage? The answers to these questions will dictate the degree to which resources are expended to control the hazard.

(b) Determination of critical control points (CCPs)

A CCP is a location, practice, procedure, or process at which control can be exercised over one or more factors so that the hazard is minimized or prevented. Two types of CCP are identified: CCP1, which will assure control of a hazard, and CCP2, which will minimize a hazard but cannot assure its control (Figure 9.1).

(c) Specification of criteria that indicate whether an operation is under control at a particular CCP

Criteria are specified limits of characteristics of a physical (e.g. time or temperature), chemical (e.g. salt or acetic acid) or biological (e.g. sensory or microbiological) nature.

(d) Establishment and implementation of procedure(s) to monitor each CCP

Monitoring is the checking that a processing or handling procedure at each CCP is properly carried out and is under control. The five main types of monitoring are visual observation, sensory evaluation, physical measurements, chemical

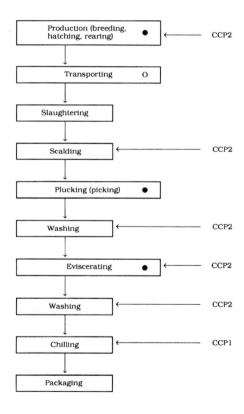

Figure 9.1 A HACCP flow chart for raw poultry meat.

testing and microbiology examination. Because the effectiveness of monitoring in terms of CCPs is related directly to the speed with which results are obtained, visual observations are often the most useful. These commonly consist of visual examination of raw materials, cleanliness of plant and equipment, worker hygiene, processing procedures, storage and transportation facilities. Monitoring procedures chosen must enable action to be taken to rectify an out-of-control situation, either before the start-up, or during the operation of a process.

(e)   Corrective action

This involves taking whatever corrective action is necessary when the monitoring results indicate that a particular CCP is not under control.

(f)   Verification

Verification is the use of supplementary information to ensure that the HACCP system is working. When verifying an established system, additional tests may be applied at a CCP that are of a more vigorous or searching nature, but not suitable for routine monitoring because of time constraints. An example would be a case where visual cleanliness is appropriate for routine monitoring of equipment hygiene. Verification that the cleaning and sanitation programme is effective and that build-up of unacceptable levels of pathogenic or spoilage micro-organisms is not occurring can only be checked by microbiological testing. End product testing for indicator organisms or pathogens is another example of verification.

Verification may be carried out by outside parties (governmental, trade or consumer organizations) and results may reveal unexpected problems. Examination of finished samples by an outside specialist laboratory may reveal hazards that were not identified in the initial hazard analysis. Reports from the market-place may indicate unanticipated public health or spoilage problems or that the product is not being handled or used in the manner expected. All these require that appropriate remedial action be undertaken and indicate a need to review the existing HACCP system.

9.4   PREDICTIVE MODELLING OF MICROBIAL GROWTH

As already noted (section 9.1), although care at harvest or slaughter may limit microbial contamination of agricultural products, the primary contamination is rarely under sufficient control to guarantee freedom from many micro-organisms of concern (e.g. *Salmonella* spp., *Cl. botulinum, Cl. perfringens, Campylobacter* spp., enteropathogenic *Escherichia coli, Listeria monocytogenes, Yersinia enterocolitica*), and with current technologies it seems unlikely ever to be so. Food processors are therefore obliged to assume that those hazardous organisms will be present occasionally in raw agricultural products, and consciously to take measures either to inactivate them or ensure that they are unable to multiply in their products.

In seeking to improve product quality and make products more desirable to consumers, food technologists are being asked to reduce heat processes, to minimize the use of even well-established preservatives, and at the same time to extend the shelf-life of foods and assure their microbiological safety. In industrially developed countries, changes in eating patterns and technological developments in food production, processing and preservation have increased the variety of products available enormously. Many are sold 'ready-to-eat' and are not, therefore, subjected to a heating process before consumption which would kill the occasional vegetative pathogenic bacteria. Some products spend weeks in distribution and storage before being offered for sale alongside the fresh product. It is now evident that the very extensive scientific literature on food microbiology lacks much information needed by the food industry to develop products that are microbiologically safe and shelf-stable.

The development of micro-organisms on foods has long been recognized to be a response to physical and chemical conditions such as pH, available water, gaseous atmosphere, temperature, preservatives and other factors. In food microbiology much effort has been directed to defining conditions that limit growth, since understanding those conditions appeared to hold out the prospect of extending shelf-life and minimizing the growth of micro-organisms associated with food poisoning. As a consequence, tables of 'minimum' values are readily available for key spoilage and pathogenic bacteria. However, many of those data have been generated when other controlling factors are near optimal, e.g. the minimum water activity when the pH value is near 7 and incubation temperatures are optimal; such pH and temperature levels are unrealistically high if the concern is food storage. Increasingly, it is being recognized that stable and safe food products are the consequence of preservative factors acting in combination often at levels which singly would not be inhibitory. Except where one or two factors limit microbial growth, our understanding of the relative contributions of the various factors to safe and shelf-stable food products is surprisingly poor.

Several years ago the Institute of Food Research recognized the need for a better understanding of the effects on growth of key micro-organisms of combinations of factors relevant to food, so that means of predicting the microbiological stability and safety of foods might be developed. The necessary database requires multi-factorial experiments, which are difficult because of inadequate bacteriological methods. The number of combinations of conditions which can be studied is limited by practical problems, such as the lack of simple automated methods for enumeration of micro-organisms, or other sensitive measures of their growth responses.

Traditionally, food microbiologists have attempted to explain the differences between the floras of the initial product and the spoiled or toxic food by analysing the initial and stored commodities for numbers and types of micro-organisms and/or their metabolites. Differences in numbers and types of organisms were then interpreted by retrospective consideration of the physical and chemical conditions during storage. Information is obtained several days after the samples are taken, and the procedures are often so lengthy as to prevent action being taken before the food is distributed. The methodologies are not suited to modern decision-making operations which may require, for example, data that can be used on-line in automated accept–reject systems.

Many of these investigations were duly reported in the scientific literature, but because experimental approaches and methodologies varied between laboratories, comparisons of the results are difficult, and data are often only relevant to the particular commodity and conditions tested. It became evident that it is impracticable to carry out microbiological evaluation of the wide, and ever-changing, range of products, processes and storage conditions currently in the market-place using traditional methods for all the bacteria of current concern. Additionally, such data would not necessarily be relevant and applicable to future products, processes and storage condition.

In consequence, the concept of 'predictive microbiology' was proposed, in which growth responses of micro-organisms of concern in foods would be modelled with respect to the main controlling factors – initially temperature, pH and a measure of available water such as water activity. The effects of additional controlling factors on growth would then be built on to that database. If models relevant to broad categories of foods could be developed, it would greatly reduce the need for *ad hoc* microbiological examination of new food products, and enable predictions of shelf-life and safety to be made speedily via a computerized database, with considerable financial benefit in the long term.

9.4.1 Probabilistic models

Where the concern is toxin production, models that predict the probability of a microbial response under given conditions are appropriate. Models have been developed for proteolytic strains of *Cl. botulinum* types *A* and *B* grown in a simulated meat product, to predict the probability of toxin production from spores as a function of NaCl, sodium nitrite (input), heat treatment, the presence or absence of other preservatives and additives such as iso-ascorbate, polyphosphate or nitrate, and incubation (storage) temperature. Many factors were tested but a few single factors (nitrite, incubation temperature, iso-ascorbate) were found to be much more important in preventing toxin production than the others, and than factors acting in combination. Table 9.7 illustrates the effects on the predicted probability of toxin production of reducing sodium chloride concentration. A comparison of data generated in the UK with similar data from the USA showed reassuringly similar trends. Subsequently, other probabilistic models or models of 'time to toxin production', have been produced for non-proteolytic psychrotrophic strains of *Cl. botulinum*; these also take account of the effect of organic acids used as preservatives.

9.4.2 Kinetic models

A different type of model has been proposed for other food poisoning micro-organisms and spoilage organisms, so that estimates of the lag (pre-exponential phase of growth during which the cells adjust to the new environment [this concept is important in our new approach to modelling]) and generation (section 3.4.1) times can be calculated. The system was developed by modelling the growth responses of three salmonellae in a laboratory medium adjusted to a range of NaCl concentrations and pH levels, and stored at a range of temperatures. The data collection proved highly labour intensive if the quality and quantity of data were to be ensured.

From individual growth curves for a wide range of combinations of pH, salt (%) and temperature, estimates of lag and growth rate were derived and modelled with respect to salt, pH and temperature via a quadratic polynomial. From the model, predictions of growth rate can be obtained for any combination of the controlling factors within the limits of experimentation. Average lag times can also be predicted but the duration of lag depends upon the physiological state of the cells which in turn depends upon their previous history and the difference between the previous and new environments. Abrupt change of pH, water activity or temperature increases the lag time. Table 9.8 compares predicted generation times of *L. monocytogenes*, as influenced by sodium chloride and nitrite concentrations, pH and temperature with

Table 9.7 Effect of reducing salt level on probability % of toxin production by *Clostridium botulinum** types A and B in pasteurized pork slurry[†]

Heat treatment[‡]		Low		High	
pH[§]		6.0	6.5	6.0	6.5
NaCl (%w/v)	NaNO₂ (μg/g)				
4.5	100	9	71	9	37
3.5	100	35	92	27	73
2.5	100	76	96	59	86*

*Mixed spore inoculum: five strains type A, five strains type B, 0.3 spores/g slurry, (10 spores/28 g bottle).
[†] Stored at 20C for up to 6 months
[‡] Low = centre temperature to 70°C.
High = centre temperature maintained at 70°C for 1 hour.
[§]Mean pH level
(Source: Robinson A., Gibson, A.M. and Roberts, T.A. (1982) *Journal of Food Technology*, **17**, 727–44.)

values taken from the scientific literature. This information can be used to model growth responses under fluctuating temperatures, such as would occur during the cooling of food items after cooking, or during distribution and display in chill cabinets.

9.4.3 Advantages and disadvantages of predictive models

The greatest advantage of predictive models is that they can be used to test the consequences of changing a number of factors at the same time and, with the power of modern computers, the answers are provided almost instantaneously. A serious disadvantage is that the prediction may not be precise and may only indicate a trend; however, knowing that trend quickly is highly advantageous when reformulating, modifying or evaluating storage conditions. Care also needs to be taken that the controlling factors included in the model are those relevant to foods in question.

Initially, there was suspicion that models generated experimentally in laboratory media would not reflect an organism's behaviour in a food. With experience, and by comparison with values for generation times published in the literature, we shall learn in which food categories growth responses in broths are mimicked, and which types of foods are exceptions to the rule. Until that experience accumulates, models should be validated in different foods before reliance is placed on them. The similarity of predicted growth parameters for salmonellae across a range of concentrations of sodium chloride, water activity, pH values and incubation temperatures to values reported in the scientific literature initially suggested that modelling might offer a cost-effective approach to understanding and controlling microbial growth in foods. Subsequently, experience with models for a range of organisms, including *L. monocytogenes*, *E. coli* 0157:H7, *Staphylococcus aureus*, and *Y. enterocolitica* has strengthened that conviction. A research programme to develop models for food-borne pathogens is being supported in the UK by the Ministry of Agriculture, Fisheries and Food (MAFF) and commercial PC-based software running under Windows™ is available as Food MicroModel. This brings the concept of predictive micro-

Table 9.8 *Listeria monocytogenes* generation times

Food	NaCl (%w/v)	pH	Temperature (°C)	Doubling time (hours)	
				Reported	IFR model predicted
Minced meat	0.0	5.5	4	28.8	28.4
Lean meat	0.0	5.6	5	22.9	22.0
Camembert cheese	2.4	6.1	6	18.0	14.4
Chocolate milk	0.0	6.4	8	10.8	9.0
Whole milk	0.0	6.2	10	6.6	6.6
11% non-fat milk solids	0.0	6.2	22	1.8	1.3

biology within the reach of large and small food companies, regulators and educational establishments. Efforts are continuing to establish international collaboration to extend models, to avoid duplication of research effort and to begin to harmonize approaches to food microbiology. In the European Community coordination under the Food-Linked Agro-Industrial Research (FLAIR) programme encouraged similar activities in member states.

Validated models can be used to judge the need for particular storage temperature, in HACCP and in beginning to quantify risk assessments.

Given validated models for the key micro-organisms associated with food poisoning and eventually with spoilage, product formulation and packaging and storage conditions could be designed to provide a microbiologically safe product of defined shelf-life. Initially, kinetic modelling has been used on viable counts (a traditional but time-consuming way of counting bacteria), but alternative measures of growth will lend themselves to automated measurement, thereby speeding the production of the models needed, to help ensure that new food products remain microbiologically safe.

10 Food safety and food legislation

A.R. Eley and I. Fisher

Food poisoning is a major public health concern worldwide. As we have shown in Chapters 2–6, it usually presents as an acute enteric disease, though there may be other forms of presentation. Acute enteric disease is usually preventable, and, as demonstrated in Chapter 9, knowledge of good microbiological practices at all stages of the food production process is an essential part of our attempts to combat food poisoning by improving food safety. Food safety has been a very topical subject recently, particularly in 1989, when there was great public concern in the UK over salmonella, listeria and botulism 'scares'. National feelings about food safety culminated in the recognition of the need for stricter controls and new legislation, which will provide greater powers for factory inspection, better training in food hygiene, and compulsory registration of food premises. This chapter gives a brief outline of some important recent developments in the fields of food safety and legislation, which it is to be hoped will result in significant improvements for the consumer.

10.1 THE RICHMOND REPORT

In 1989, in response to a series of well-publicized food safety concerns, the UK government announced the setting up of a Committee on the Microbiological Safety of Food, to be chaired by Sir Mark Richmond; this Committee's findings have subsequently become known as the Richmond Report, Parts I and II. The Committee was given the following broad terms of reference:

'To advise the Secretary of State for Health, the Minister of Agriculture, Fisheries and Food and the Secretaries of State for Scotland and Northern Ireland on matters remitted to it by Ministers relating to the microbiological safety of food and on such matters as it considers need investigation'.

More particularly, the members of the Committee were asked to look at specific questions relating to the increasing incidence of food poisoning in the UK, especially that associated with salmonella, listeria and campylobacter; to establish whether this was linked to changes in agriculture, food production, food technology, retailing, catering and food handling in the home, and to recommend action where appropriate.

10.1.1 Major recommendations of Part I

Part I of the Richmond Report was published in 1990. It covered public health matters including the incidence of food poisoning, human and food surveillance and outbreak management, as well as poultry meat production and manufacturing processes. As might be expected, this part of the report contained many significant conclusions and recommendations; the recommendations listed below are those which we felt to be of greatest overall importance:

1. Further research on the natural history of listeriosis and the factors influencing the growth of listeria in different foods, so that preventive measures can be determined.
2. More studies aimed at finding the 'true' incidence of gastrointestinal illness, the related microbial cause(s) and possible food source(s).
3. Closer liaison between the Communicable Disease Surveillance Centre (CDSC) with data on human cases and the State Veterinary Service (SVS) which provides data on animal diseases. This might identify links between contaminated animal foods and the same pathogens causing human disease.
4. The appointment of Consultants in Communicable Disease Control (CCDCs) by every District Health Authority to take overall responsibility for outbreak management.
5. Further training of all relevant staff in the practical management of local and national outbreaks and all aspects of food safety.
6. More coordinated food surveillance studies at centres throughout the country, to provide national data on the microbiological quality of food.
7. A microbiological monitoring programme to investigate the level of contamination at each stage in poultry-meat production.
8. Further training of staff involved with food production, including the seeking of expert advice on food processes, which should be designed on HACCP principles.
9. The initiation of a registration or a licensing procedure which will include the inspection and approval of new premises for the production and/or sale of food.
10. The replacement of the present 'sell-by' labelling system of food by a uniform 'use-by' system. This will have updated requirements and guidance to help determine a safe 'use-by' date.

10.1.2 Major recommendations of Part II

Part II of the report, published in 1991, included an overview of the causes of food poisoning, and identified changes responsible for its increased incidence.

Detailed consideration was given to the microbiological hazards associated with red meat, milk, cheese and shellfish, and the complexities of retail operations were analysed. It was suggested that HACCP should be implemented across the entire food industry, and that greater emphasis should be placed on education and training. The specific recommendations included:

1. The need to work towards a common definition of food poisoning not only for all countries of the UK but for the whole of the European Community.
2. An appeal for harmonization of reporting of food poisoning statistics across the UK, while recognizing the public health arrangements peculiar to Scotland and Northern Ireland.
3. Training of meat inspectors, Environmental Health Officers and veterinarians once new EC arrangements for meat inspection are agreed.
4. In the dairy industry, advice on the care required in cheese making with respect to listeriosis, following the implication of the consumption of unpasteurized milk as a major hazard.
5. An urgent need to look at the contamination of shellfish beds following the failure of depuration (cleansing by subjecting them to a continuous flow of decontaminated recycled salt water) to remove enteric viruses.
6. The setting up of a Code of Practice based on existing guidelines for all transport of refrigerated foodstuffs, to include vehicle monitoring facilities and adequate training of staff handling refrigerated foods.
7. The undertaking of new initiatives to review all microbiological safety aspects of retailing and wholesaling.
8. The maintenance of high standards in the planning and operating practices in catering, with special attention being given to HACCP guide-lines.
9. The provision of advice to consumers on the optimal utilization of domestic refrigerators and microwave ovens.
10. The setting up of major new initiatives for improved education and training in food hygiene and food safety issues, related research to be encouraged nationally and with international collaboration.

If all the recommendations outlined in sections 10.1.1 and 10.1.2 above are implemented, they should have a significant impact on the level of microbiological contamination of food at all stages from production to consumption. Over a period of time this should result in fewer cases of food poisoning in this country. The subsequent introduction of new legislation relating to food safety and standards of hygiene in the food industry should provide a legal framework for the enforcement of a number of the Richmond Report's main conclusions.

10.2 UK FOOD SAFETY LEGISLATION

The origin of legislative control of the food supply lies in the need to prevent fraud and adulteration as a potential risk to the health of the consumer (and as a means of making large profits). The development of the mass market and the experiences of the emergency legislation which controlled the shortages, composition, labelling and pricing of food during the Second World War led to the introduction of the Food and Drugs Act 1955. This was superseded by the Food Act 1984, which set out certain rules relating to the purity and quality of food and the hygiene standards to be observed, and provided a maximum penalty of £2000 for selling food which was not of the quality demanded. Apart from this, however, there had been little movement in primary food legislation in the last 30 years; yet during this time there have been major changes in the way in which food is processed, prepared and sold. It is not surprising that the many food scares of the late 1980s (particularly salmonella in eggs), have led to new provisions in the Food Safety Act 1990, which came into effect on 1st January, 1991. Moreover, the new powers for enforcement authorities were strengthened by the Food Hygiene (amendment) Regulations 1990 which required certain chilled foods to be stored at 8°C or 5°C, although it is unclear as to whether or not the 5°C was ever enforced as it proved difficult for the industry to comply with consistently.

It had been known for several years that these powers for enforcement authorities would have to be replaced by legislation that met with the demands of the EC Directive on the Hygiene of Foodstuffs, (93/43/EEC) the Food Hygiene (General) Regulations and many others; these were eventually superseded by the Food Safety (General Food Hygiene) Regulations 1995 on September 15th 1995. The safety of food, its means of transport, any processing it might undertake and the premises in which it is handled is now statutorily controlled more comprehensively than ever before.

The Government is committed to reducing the burden of regulation on businesses, to reducing unnecessary red tape and to educating inspectors to apply a common-sense approach to enforcement. Some 450 changes to legislation are to be made, and all new law must take into account the needs of small businesses and the benefits to be gained by regulation. In the food industry sector there are several proposals, from a review of the need to register food premises to changes in the food temperature control rules.

It is clear that with the conflicting demands of EU Directives and deregulation, detailed legislation to control the industry will be changing for some years to come.

10.2.1 The Food Safety Act 1990

Local Authority enforcement officers (Environmental Health Officers or Trading Standards Officers) have been given several new powers under this piece of legislation that will greatly enhance their ability to deal with contraventions of the law quickly and efficiently. The discussion which follows gives an indication of the major provisions of the Act, though it is by no means an exhaustive resumé. Table 10.1 sets out the maximum penalties which are provided for

Table 10.1 Penalties for offences under the Food Safety Act 1990

Offence	Maximum penalty
Obstructing an enforcing officer, failure to give assistance or information	£2000 fine or 3 months' imprisonment, or both
Rendering food injurious to health (under section 7)	£2000 fine or 6 months' imprisonment, or both*
Selling food that does not comply with safety requirements (under section 8)	£20 000 fine or 6 months' imprisonment, or both*
Selling food that is not of the nature, substance or quality demanded by the purchaser (under section 14)	£20 000 fine or 6 months' imprisonment, or both*
Other offences	A fine or 6 months' imprisonment, or both*

*These penalties apply on summary conviction; for conviction on indictment, the maximum penalty in all cases is a fine or 2 years' imprisonment, or both.

under the Act, and gives some indication of how seriously food safety offences are now regarded at government level.

(a) Food safety requirements

There is now a new offence of selling, offering for sale, possessing or advertising for sale food that does not comply with a food safety requirement. Food is deemed to have failed a safety requirement if it is: (a) injurious to health; (b) unfit for human consumption; or (c) contaminated with materials that preclude its use for human consumption. Where food fails to comply with a safety requirement the whole batch is presumed to be affected until the contrary is proved.

The person responsible for the food (usually the owner of the business) is now liable for prosecution if he/she fails to ensure that it satisfies food safety requirements. This emphasizes the need for documented systems of control of foodstuffs, from the method of approval and regular auditing of suppliers to storage and subsequent processing and distribution.

(b) Food safety notices

Enforcement officers have been given very detailed and powerful new provisions for dealing with equipment, processes and premises that contravene the legislation or pose a threat to the health of the consumer. The old choice for the officer of an informal letter or of prosecution is now supported by the ability to serve *at the time of inspection* a notice on the person committing the offence. This notice may give the recipient time to correct the fault (improvement notice) or may close down immediately a piece of equipment, process or premises (prohibition notice).

The effect of these notices on a food business could be profound. Although there has been some publicity regarding inconsistency of approach by different authorities, much of this has proved to be incorrect. Local Authorities follow Codes of Practice and many activities are coordinated nationally to reduce the possibility of incorrect or inappropriate notices being served. Well-managed businesses need not normally fear visits by enforcement officers, but should be ready to respond immediately: managers need to react quickly and positively to receipt of a notice and be proactive in an attempt to limit the circumstances where a local authority would feel that service of notices is necessary.

(c) Registration of premises

It is now a requirement for all 'food businesses' to register with the Local Authority. Not to do so before opening is an offence. It is alleged that this enables Local Authorities to plan inspections and look at new businesses before they open. It is of doubtful value, but is still a requirement, although that might well change with the deregulation initiative of the Government.

Local contact with the Environmental Health Department should enable the registration of all food businesses to be carried out quickly and easily.

(d) Food hygiene training

It has long been acknowledged that the safety of food is only achieved by correct handling and storage throughout the chain of supply, from producer via supplier to the customer. At each of these stages there is the opportunity for human error, and as one of other major causes of human error is ignorance, there is a need for the adequate training of anyone who may in any way affect the safety of food. This includes food-handlers, the designers and engineers who may work with equipment involved with food, producers of food packaging, the managers of food-handlers and many others.

For many years, organizations such as the Chartered Institute of Environmental Health, the Royal Society of Health and the Royal Institute of Public Health and Hygiene have run certificated training courses in food hygiene for food handlers, supervisors and managers (see Table 11.2). For many in the food industry, such as caterers, these courses are to be recommended but we must be aware that general courses do not always satisfy a particular training need. This has been recognized in the new Food Safety (General Food Hygiene) Regulations 1995, which require the proprietor of a food business to 'ensure that food handlers engaged in the food business are supervised and instructed and/or trained in food hygiene matters commensurate with their work activities'. This acknowledges that there is a relationship between the procedures adopted by the business, and the supervision of those procedures and instructions being followed by the food-handler after training. It is therefore recommended that food businesses carefully consider the training needs of the employees, how the training is provided, and how the effectiveness of the training is evaluated.

10.2.2 Food Safety (General Food Hygiene) Regulations 1995

These Regulations became law in September 1995, replacing or amending 17 previous sets of regulations, perhaps most importantly the Food Hygiene (General) Regulations 1970. They do not apply to those parts of the food industry covered by product specific directives such as meat, meat products, fish and eggs. The importance of the regulations to the food industry cannot be over-emphasized, as they now set the standard of food control expected of the industry and allow for heavy penalties if the standard is not achieved.

The regulations are few and brief, and are given detail by the 10 chapters of Schedule 1 (see below).

The major implications for the food industry are as follows:

(a)   Regulation 2 – There is now a statutory definition of 'Hygiene': 'all measures necessary to ensure the safety and wholesomeness of food during preparation, processing, manufacturing, packaging, storing, transportation, distribution, handling and offering for sale or supply to the consumer'.

(b)   Regulation 4 – There is now a clear requirement for the proprietor of a food business to critically assess the operations of the business and take steps to control the safety of food. The regulation does not mention HACCP or Assured Safe Catering systems, but does say: 'a proprietor of a food business shall ensure that any of the following operations, namely, the preparation, processing, manufacturing, packaging, storing, transportation, distribution, handling, and offering for sale or supply, of food are carried out in a hygienic way', and carries on to require the proprietor to '. . . identify any step in the activities of the food business which is critical to ensuring food safety and ensure that adequate safety procedures are identified, implemented, maintained and reviewed . . .'.

(c)   Regulation 5 – This requires persons working in food handling areas to report to the proprietor if they know or suspect that they are suffering from or are the carrier of any disease likely to be transmitted through food; or if they have an infected wound, skin condition, sores or diarrhoea. The proprietor must then decide if the person is to be allowed to work with food or within the food handling area.

(d)   Regulation 8 – This requires food authorities to inspect premises regularly and will depend on the risk involved within the business, i.e. higher risk operations will be inspected more frequently than less risky businesses. This regulation also refers to codes of good hygiene practice which have been forwarded by the Government to the European Commission.

Guides will be originated by the industry concerned and when completed submitted to the Government for recognition. Where a business is following advice in a recognized guide, this must be given due recognition by the enforcement agencies. Recognized guides can be used with confidence by businesses as a practical guide to compliance with the relevant regulations. It will not be a legal requirement to follow the guides and it will be possible to convince enforcement officers that other methods achieve at least the objective of the guide.

Industry guides to good hygiene practice will be accepted by the Government if drawn up by the industry and subject to wide consultation. It is almost inevitable that these will become the standards of food hygiene enforced in the industry.

Schedule 1 – This is broken down into 10 chapters and covers:

(i)   general requirements – cleanliness, layout, design, size, facilities, etc.
(ii)   structure – walls, floors, ceilings, windows, doors, surfaces, etc.
(iii)   movable and temporary premises, domestic premises and vending machines design, facilities, cleaning, etc.
(iv)   transport – cleanliness, design, prevention of contamination
(v)   equipment – construction, repair, installation, etc.
(vi)   food waste – segregation, collection and removal
(vii)   water supply – adequate and potable water, including ice and steam
(viii) personal hygiene – personal cleanliness, protective clothing
(ix)   food – segregation of unfit food, prevention of cross-contamination
(x)   training – the proprietor ... shall ensure that food handlers are
        - supervised and
        - instructed and/or
        - trained
        in food hygiene matters commensurate with their work activity.

The schedule does not give precise and detailed standards for food premises, but does give general requirements. Specific matters will be covered in the industry codes (see above). The only major change from the 1970 regulations is chapter x, requiring planned provision of training and instruction together with supervision, commensurate with the work activity.

10.2.3 Food Safety (Temperature Control) Regulations 1995

As it proved difficult to achieve agreement within the European Union for the UK standards on the temperature control of food, the new general regulations did not include these provisions. They did, however, become law on the same day and again set important standards to be followed by the industry.

The major change from the 1990 amendment regulations is that there is now no reference to 5°C at all. Chilled food is required to be held at <8°C and hot food at >63°C with defences regarding time limitations and manufacturers' guidance. There is now the long awaited additional requirement to cool hot food as quickly as possible, and an interesting extra regulation which says:

'. . . no person . . . shall keep foodstuffs . . . at temperatures which would result in a risk to health'. This reflects the fact that some pathogens do grow below the statutory 8°C and appears to indicate that even if the proprietor complies with the regulations, if food poisoning results then they have probably committed an offence.

10.2.4 The due diligence defence

A new provision brought into food safety legislation is the defence of 'due diligence'. This defence is available to individuals and organizations that are being prosecuted by the local authority enforcement officers (EHOs or TSOs).

The existence of a comprehensive and efficiently operated due diligence system has the potential to render those being prosecuted immune from prosecution. Such a system must be capable of withstanding the strictest and most critical scrutiny of enforcement officers. It will lead to acquittal, even if the facts show that an offence has occurred.

There is now available the defence in section 21 of the Food Safety Act 1990. 'It should be a defence for a person charged to prove that he took all reasonable precautions and exercised all due diligence to avoid the commission of the offence by himself or by a person under his control'.

It is clear that to avoid prosecution, it will be necessary to prove that the requirements of the full defence have been fulfilled. The prosecution is under no obligation to prove otherwise. The burden of proof is on balance of probability, i.e. less than 'beyond reasonable doubt', but proof will be required in documentary form. If, for any reason, *all* reasonable precautions have not been taken, the defence will fail.

It is notable that the principles involved in good management, as reflected by Quality Assurance and BS5750 (ISO 9000) are recognized as being the basis of a good due diligence defence. However, it must be understood that being accredited under the scheme will not necessarily mean that the defence will be successful.

10.3 CONCLUSIONS

One of the major factors that led to the Government's action and resulted in the publication of the Richmond Report Parts I and II and the new legislation was the increased incidence of food poisoning in the UK. In some ways these developments were also timely, firstly as the 1984 Food Act was in need of updating, and secondly, so that we could implement EC food law in the single market (1992).

As technology advances and public awareness grows, so consumers are becoming increasingly demanding in terms of the choice, quality, freshness, nutritional value and microbiological safety of food. The current emphasis on 'healthy' eating will no doubt mean a continuing decline in the use of preservatives in food, a potentially hazardous trend, since preservatives are often directly responsible for reducing the levels of food poisoning bacteria. New ways of preserving food will therefore be needed. One approach to the problem may be through the use of synergistic combinations, for example moderate heating of a product followed by cold temperature storage. Moreover, as our eating habits have changed markedly over the last 20 years (as shown by our desire for more exotic food), and as international travel over this time-period has greatly increased, so the possibilities for food-borne spread of infection have also increased. Fortunately, the same time-span has also brought advances in food technology to help curtail opportunities for microbiological hazards, as well as significant developments in laboratory diagnosis, for example, novel techniques such as the polymerase chain reaction (PCR). These new scientific advances will give us the possibility not only of detecting pathogens much more quickly, but also at a much higher level of sensitivity. It is to be hoped that we will now be able to determine a greater percentage of aetiological agents, which are sure to include viruses in increasing numbers, in those outbreaks of food poisoning previously recorded as of unknown aetiology. We must appreciate, however, that conclusive proof that organisms are positively implicated in the causation of food poisoning can be difficult to obtain; for unusual organisms, those found in small numbers or those of equivocal virulence potential, problems will only increase.

Although there are many factors to consider, and recent trends in the incidence of food poisoning have given considerable cause for concern, the overall impact of the developments outlined in this chapter should lead in time to a major improvement in the microbiological safety of food.

11 Food hygiene

A.R. Eley

Hygiene can be defined as knowledge or practice relating to the maintenance of health, or a system of principles for preserving or promoting health. In relation to food, hygiene therefore implies the observance of principles and practice which minimize the risk of food becoming a vehicle for the transmission of pathogenic micro-organisms. Food hygiene relates to the safety of food at all the stages which lead up to its consumption: from production through processing and storage to the circumstances in which it is eaten.

Essentially, in order to keep food safe, we need to try in the first place to prevent contaminating organisms from coming in contact with it. However, this is often impracticable: some foods, such as poultry, are often contaminated at source, while others are necessarily exposed to the risk of contamination during processing. In these circumstances, good food hygiene practice means limiting multiplication of contaminating organisms to as great a degree as possible by controlling the conditions under which food is processed and stored. Finally, we can aim to destroy harmful bacteria by appropriate cooking methods. In all this, we can identify three key factors – the personnel involved, the nature of the foodstuffs themselves, and the food preparation environment – whose importance for food hygiene and food safety are explored below.

11.1 PERSONNEL

All food-handlers, be they commercial or domestic, have a responsibility to ensure that their actions do not lead to food poisoning organisms being introduced into food or, when they occur naturally, to their being encouraged to multiply so as to cause disease.

11.1.1 Personal hygiene

Essentially all food-handlers should be aware that personal hygiene is of the utmost importance, as any minor breakdown could directly lead to food contamination. They also need to realize that even in health the human body has its own microbial flora on the skin, face, nose and hair which often includes pathogens such as *Staph. aureus*. It is therefore important that physical contact with food be reduced as much as possible to limit the incidence of contamination. Food-handlers should be encouraged to keep their hair clean and wear some sort of head covering, and to avoid touching the face, particularly the nose as much as possible (Table 11.1).

Hands are probably the principal route for the transmission of pathogens to food and, as has been described elsewhere in this text, handling should be reduced to the essential minimum. However, it is obvious that some handling of food has to take place, so, when this is necessary, hands and nails should be as clean as possible. This means keeping nails short, and thorough hand washing on a regular basis. This is especially true when in contact with heavily contaminated products such as raw meat and poultry; it should also be normal practice to wash the hands between handling raw and cooked foods. Particular care should be exercised in relation to anyone who has a cut or other wound on the hand. Uninfected wounds should be covered with waterproof dressings to prevent their contamination and resultant infection by *Staph. aureus* present on the skin. Anyone who has an infected wound or boil on any part of the body should not be allowed to handle food, because of the risk of transferring the organism in large numbers from the infected area to the hands and thence to foodstuffs.

It should always be standard practice to wash the hands after visiting the toilet, in order to prevent faecal contamination. It should be remembered that even normal faeces may contain as many as 10^{12} bacteria per gram. Not surprisingly, with this large number of organisms, food can easily be contaminated if there is any lapse in hygiene. This is particularly important in the case of certain bacteria such as *Campylobacter* or *Shigella* that only require a low

Table 11.1 Habit categorization of food-handlers

Desirable	Undesirable
Regular attendance at food hygiene courses to update knowledge	Only Basic Food Hygiene Certificate obtained
Early reporting of food poisoning symptoms	Symptoms ignored: personnel continue to work
Clean hands with short finger nails	Not washing hands, especially after visiting the toilet or handling raw food
Covering skin cuts using a sterile waterproof dressing	Allowing an infected wound to contaminate food
Wearing of clean protective clothing	Wearing of jewellery
Not smoking, coughing or sneezing near	Smoking, coughing or sneezing near food
Clean hair with a head covering	Unclean hair and/or not wearing a head covering

infective dose to cause disease. This is the reason why any person displaying symptoms of food poisoning or any other enteric disease, or who is thought to be a contact of a confirmed case should not be allowed to handle food. As a general rule any person known to be excreting a food poisoning organism should not be allowed to handle food until they are symptom free and have tested negative in three consecutive samples of faeces.

11.1.2 Training and education

From a commercial point of view there have been some new developments in training and education since the introduction of the Food Safety Act 1990. This Act empowers Ministers to make regulations requiring the hygiene training of persons engaged in food businesses, and Food Authorities to provide food hygiene training courses for such persons. One way for organizations to comply with proposed legal requirements for training is to establish a planned training programme. This can be in-house training or training by reputable and experienced outside bodies. Such bodies can provide a number of courses depending on the level of training required, and can award a nationally recognized qualification (Table 11.2).

All commercial food-handlers should be given basic training which includes personal hygiene, principles of hygienic work practice to prevent contamination

Table 11.2 National food hygiene courses

Body	Course
Institute of Environmental Health Officers	Basic Food Hygiene Certificate Intermediate Food Hygiene Certificate Advanced Food Hygiene Certificate
Royal Environmental Institute of Scotland	Elementary Food Hygiene Course Intermediate Food Hygiene Course Diploma in Advanced Food Hygiene
Royal Institute of Public Health and Hygiene	Primary Certificate in Hygiene for Food Handlers Certificate in Food Hygiene and the Handling of Food Diploma in Food Hygiene Diploma in Hygiene Management
Royal Society of Health	Certificate in Essential Food Hygiene Certificate in Hygiene of Food Retailing and Catering Diploma in Food Hygiene Management
The Society of Food Hygiene Technology	Hygiene Training Scheme for Food Handlers in Food Manufacturing and Supplying Industries

and food poisoning, maintenance and cleaning of premises and equipment, and food hygiene legislation. Further training will be dependent on the risk to food safety posed by the tasks undertaken together with the available level of supervision. Moreover, as the HACCP approach to food safety is now more widely accepted, food-handlers should be aware of its importance in relation to fulfilling responsibilities of the critical control points (p. 169).

However, we should remember that out of the reported general and family outbreaks of food poisoning in England and Wales, as many as 25% may be caused by food prepared in the home. There is an urgent need to improve the food hygiene standards of the general public as well as those of commercial food-handlers. Lack of knowledge about food safety has been identified as the major factor in poor hygiene in the home. It may be time to reconsider how the general public are educated in food safety and hygiene. A variety of different types of educational materials are already available from a number of sources such as educational establishments; news media; storage and cooking instructions; leaflets and other forms of advertising material; advice from family, friends, EHOs, GPs, etc. However, more specific guidance may be needed on particular problems such as temperature control, proper use of kitchen appliances (for example microwaves) and correct handling of new food products. In addition, if the public are taught about the basic and most important factors that contribute to food poisoning, this may help to promote a better understanding of food hygiene. Guidelines for the safe handling of foodstuffs are more likely to be widely accepted and followed if people have some knowledge about the likely effects of poor hygiene.

11.2 FOODSTUFFS

We are primarily concerned here with ways of trying to reduce the hazards associated with contaminated food. One way of reducing these hazards is to prevent or slow down microbial growth, which is itself dependent on a number of factors such as nutritional content of the food, temperature, pH, gaseous conditions, presence of inhibitors, and water activity. All these factors are well known to food microbiologists, who exploit them for food preservation, which is the treatment of food to prevent or delay spoilage and inhibit growth of pathogenic organisms which would render the food unfit. Where food hygiene is concerned, two factors are particularly important, namely water activity and temperature.

Water activity can be defined as the water requirement needed for microbial growth (or the level of dehydration). As a guideline, the a_w or amount of available water for pure water is 1.00. Most bacteria cannot grow at an a_w below 0.90; many yeasts and moulds can grow at an a_w below 0.90 but not usually below 0.80, whereas xerophilic fungi and osmophilic yeasts can grow at an a_w of as low as 0.65. These differential a_w requirements of micro-organisms give us a scientific basis for risk categorization of foods.

As most food poisoning is caused by bacteria, foods can be divided into those that will support their growth, i.e. where the a_w is above 0.90, and those that will not. From this point of view, the following foods can be termed potentially dangerous:

- fresh meat
- fresh seafood
- fresh vegetables
- milk
- most cheeses

The following foods with an a_w of below 0.90 can be spoiled by organisms such as yeasts and moulds, etc. but can be termed potentially safe with regard to the presence of food poisoning bacteria:

- flour
- biscuits
- dried eggs, milk and vegetables
- jam
- dried fruit
- nuts

However, these foods can occasionally be hazardous, or even fatal: hazelnuts were responsible for the botulism outbreak in the north-west of England in 1989.

Bacterial growth in potentially dangerous foodstuffs is very significantly affected by temperature. Recently, there has been a great emphasis on proper refrigeration, together with strict temperature control as a means of reducing the risk of food poisoning. It needs to be stressed, however, that refrigeration is a means of delaying and not preventing food spoilage, and that although most food poisoning bacteria are not able to grow at 4°C, a number are, such as *L. monocytogenes*, *Y. enterocolitica*, *Aeromonas* spp. and some strains of *Cl. botulinum*. Generally speaking, the cooler the refrigerator temperature the lower the rate of organism growth, which is why the maximum temperature for commercial storage of most high-risk foods has been set at 5°C. Length of storage is also an important factor, even at 4°C the organisms mentioned above may multiply sufficiently over a period of a few days to cause disease. This explains why no maximum temperature has been set for storage of sandwiches intended for sale within 4 hours of preparation, while those intended for sale between 4 and 24 hours must be kept at no more than 8°C, and those which will be kept for more than 24 hours before sale are subject to a maximum storage temperature of 5°C (Table 11.3).

From a domestic viewpoint, emphasis has been placed on the maintenance of a cold chain from purchasing products in the store to placing them in the refrigerator/freezer at home. Recommendations include trying to buy chilled/

Table 11.3 The statutory maximum temperature for the cold storage of relevant food

Relevant food	Temperature (°C)*
Soft ripened cheeses which have been cut or otherwise separated from the whole cheese, for example, Brie, Danish Blue, Camembert. Hard and soft cheese included in a cooked product	5
Cooked products containing meat, fish, eggs (or substitutes), cheese, cereals, pulses or vegetables, which are intended for consumption without further reheating, for example, cold cooked meats, meat removed from the can, cooked vegetables or cereals, salads, meat and fish pâté, scotch eggs, pork pies with gelatine added after cooking, quiches, sandwich fillings, e.g. chicken, egg mayonnaise or tuna	5
Smoked or cured fish whether whole or cut after smoking or curing	5
Smoked or cured meat when cut/sliced after smoking or curing, for example, cured cooked hams, salamis and other fermented sausages	5
Prepared vegetable salads containing high-risk relevant foods such as pasta salad	5
Sandwiches, filled rolls and bread products containing soft cheese, smoked or cured fish or meat, cooked products, as above, and not intended to be sold within 24 hours	5
Any sandwich intended to be sold within 4 hours of preparation	Not covered
Sandwiches containing any other relevant foods and any sandwich intended to be sold within 4 to 24 hours of preparation	8
All other relevant foods, unless specifically excluded, for example, whole ripened soft cheeses; cooked products intended for further heating before eating, for example, certain meat pies, pizzas and many ready-made meals; dairy-based desserts with a pH value of 4.5 or more, for example, most trifles, creme caramels, whipped cream desserts; prepared vegetable salads, for example, coleslaw; uncooked or partly cooked dough products containing meat or fish, for example, pizzas; fresh pasta with meat or fish fillings, for example, ravioli; dairy creamcakes.	8

*Refers to the temperature of the food, not the air temperature.
(Source: Sprenger, R.A. (1993), *Hygiene for Management: A text for food hygiene courses*, p. 239.)

frozen foods last and keeping them together to prevent them heating up too rapidly. Consumers are also advised to try to take such products home within an hour or so (or make use of cool bags or boxes) and once home to store them away quickly in the refrigerator/freezer. The temperature of a domestic refrigerator should be between 0 and 4°C and that of a freezer −18°C. The only way to check that these temperatures are being maintained is to use a fridge/freezer thermometer. Temperature control is particularly important where chilled convenience foods are concerned. These require careful storage, both in retail outlets and at home, and special attention to shelf-life if food poisoning is to be avoided.

The increased consumption of chilled convenience foods is just one example of changes in lifestyle which pose a potential threat to food safety. Many consumers now shop only once a week (or even less frequently for certain major items), which means a greater tendency to store foods for longer. There is also currently a trend for healthy foods, especially those that do not contain preservatives or microbial growth inhibitors. Whereas previously products containing preservatives might not have required strict temperature control, the healthy alternatives are likely to be highly perishable, and so require refrigeration. This in term means that greater demands will be placed on refrigerator space, and it is therefore important that this space be used as safely as possible. In order to encourage cold air circulation there should be no overcrowding of shelves. It should always be remembered that if foods need to be stored in a refrigerator or freezer after cooking, they should be cooled rapidly before being refrigerated/frozen. As it is not advisable either to place hot food directly into a refrigerator or to leave it cool at room temperature for a considerable period of time (which might encourage spore germination and bacterial growth), hot food should be cooled as quickly as possible, by placing it in a cold room such as a pantry, by diluting it with cold liquid or by submerging in cold water.

When it comes to cooking food items taken from the freezer, many can be cooked immediately without defrosting, providing that cooking times are increased accordingly. The exceptions are poultry and other large items, which must be completely thawed before cooking. If food is not completely thawed at the centre before cooking, then heat will often only melt the ice and not raise the internal temperature above that required to destroy pathogens.

For keeping cooked food hot, temperatures must be high enough to prevent bacterial growth, which means 63°C or above. Figure 11.1 clearly illustrates which temperatures are safer for storage of food. If these are not observed, and the temperature achieved is the optimum for bacterial growth, then it is easy to understand the dramatic rise in numbers of bacteria which could ensue (Figure 11.2) and lead to a disastrous consequence.

11.3 ENVIRONMENT

The environment for food preparation is also important in trying to prevent microbial contamination of food. All surfaces with which food may come in contact should be carefully cleaned to remove matter conducive to the growth of bacteria: splashes of soup or gravy on tiles or working surfaces and crumbs lodged in cracks and corners are potential breeding grounds for organisms. Removal of such matter followed by use of a disinfectant such as hypochlorite is a good combination treatment. Disinfection is also recommended for utensils and other kitchen equipment. It should be remembered that there is a difference between sterilization and disinfection, and the two

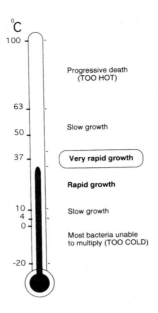

°C

100 — Progressive death (TOO HOT)

63 — 50 — Slow growth

37 — Very rapid growth

Rapid growth

10 — 4 — 0 — Slow growth

Most bacteria unable to multiply (TOO COLD)

-20 —

Figure 11.1 How food poisoning bacteria react to different temperatures. (Source: Eley, A. (1992) *Nutrition and Food Science*, **5**, 8–13.)

terms should not be confused. The term sterilization relates to the destruction of all microorganisms and their spores, whereas disinfection refers to a dramatic decrease in micro-organisms to a level which is not harmful to health; however, bacterial spores are not usually affected by disinfection.

If food preparation surfaces such as chopping boards become contaminated with food poisoning organisms, then microbial cross-contamination can easily result. This is why the same equipment cannot be used for preparing raw and high-risk products without suitable cleaning and/or disinfection. As a general rule raw food should always be handled and stored separately from cooked food. Likewise, it makes sense to keep food covered when not being handled, in order to prevent contamination from, for example, flies, other insects and accidental skin contact. Apart from microbial contamination there may also be unwanted chemicals on fresh fruit and vegetables, and therefore it is always advisable to wash them, particularly if they are to be eaten raw.

11.4 CONCLUSIONS

The discussion in this chapter has focussed on three important areas of food hygiene: the behaviour of food-handlers, the nature of foodstuffs and their implications for storage and cooling, and the design and maintenance of the food preparation environment. Underlying the discussion of all these areas

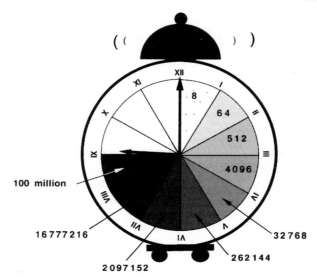

Figure 11.2 Growth rate of bacteria at their optimum temperature (numbers around the clock face in hours; numbers within the hours are bacteria). (Source: Eley, A. (1992) *Nutrition and Food Science*, **5**, 8–13.)

there is a single theme, namely the importance of education in ensuring the safety of food and minimizing the risk of food poisoning.

There are now large number of courses available, ranging from basic to advanced, on food hygiene for professionals (Table 11.2). The emphasis on this type of training has also been reinforced by the recommendations of the Food Safety Act 1990, and in theory standards of food hygiene at work should be improving. However, although efforts have been made to improve general awareness, using a range of educational materials, consumers still do not appear to be very knowledgeable about some of the essential principles of food hygiene. This was clearly demonstrated by a questionnaire administered in South Wales in 1991, when respondents were asked to comment on practices in the domestic kitchen and to give their views on food safety and food poisoning. Some of the findings of this questionnaire are as follows:

1. 77% of respondents had never measured the temperature of their refrigerator and 61% did not know what temperature range their refrigerators should be within.
2. 36% of respondents put food items in the refrigerator wherever there was space and 28% never stored raw meat on the bottom shelf of the refrigerator.
3. 25% of respondents who owned a microwave oven did not know the power of their oven and 19% never left food to stand for the recommended time after cooking or reheating in the oven.
4. 58% of respondents prepared meals in advance, either for eating later that

day or on another day; of these 37% sometimes stored the cooked meals on the kitchen work surface and 24% stored the food in a saucepan on top of the cooker.

5. 42% of respondents failed to understand that keeping food at room temperature or contamination of food after cooking could lead to food poisoning.
6. 59% of respondents thought that food could be made safe by freezing it and 26% considered that food could be made safe by storing it in a refrigerator.

Statistics like these underline the need for a continuing programme of consumer education designed to heighten awareness of the causes of food poisoning and of proper food hygiene practices.

In addition to what has been stated already in this chapter, the following preventive measures can be used as guidelines to improve kitchen hygiene:

1. Store raw food at the bottom of the refrigerator and cooked foods at the top, to help avoid the risk of cross-contamination caused, for example, by blood from raw meat dripping onto items stored below it.
2. If food is thawed at room temperature, remember to cool or cook it immediately after thawing as the warmer conditions will be ideal for organisms to multiply.
3. Avoid eating undercooked poultry and raw meat.
4. Keep food for as short a time as possible, i.e. observe 'use-by' dates.
5. Never serve food that looks or smells peculiar; on the other hand, many foods implicated in food poisoning look and taste quite normal.
6. Try to achieve a high enough internal temperature when cooking large joints or birds.
7. Never reheat food more than once.
8. As a general rule, reheating cooked food in a microwave oven is safe, but this mode of cooking is not recommended for cooking raw meats and raw meat products.
9. Some microwave ovens produce 'cold spots' in food, so a more even distribution of heat is required; this can be achieved by thorough stirring during or after cooking.
10. Keep eggs in the refrigerator and use within 3 weeks of the laying date.
11. Avoid eating lightly-cooked or raw eggs; this also applies to foods such as mousses, sauces or custards which contain raw or lightly-cooked egg whites and/or yolks.
12. Particular groups of people such as the elderly, sick, babies and pregnant women should only eat eggs cooked until both yolk and whites are solid.
13. Keep pets away from food, dishes and worktops.
14. Don't let birds drink the milk that's delivered to your door. Also, bring the milk indoors and store in the refrigerator as soon as possible.

Food safety legislation alone can only go part way towards reducing the incidence of food poisoning; it must be complemented by efforts to improve the overall standard of education among consumers. This is a fundamental area, where progress could easily be made, for example by teaching basic food hygiene in schools, or by persuading the makers of television cookery pro-grammes to include a regular food safety slot in their broadcasts. Only in this way will some of the worrying statistics outlined elsewhere in this text about the prevalence of microbial food poisoning be likely to improve.

Recommendations for further reading

The following are a number of textbooks or publications that should be useful reference material on general topics covered in this book.

Board, R.G. (1983) *A Modern Introduction to Food Microbiology*, Basic Microbiology Series, Volume 7. Blackwell Scientific Publications, Oxford.

Frazier, W.C. and Westhoff, D.C. (eds) (1988) *Food Microbiology*, 4th edn, McGraw-Hill, Singapore.

Hobbs, B.C. and Roberts, D. (1993) *Food Poisoning and Food Hygiene*, 6th edn, Edward Arnold, London.

Microbiological Safety of Food, Part I. Report of the Committee on the Microbiological Safety of Food. Chairman: Sir Mark Richmond. HM Stationery Office, London, 1990.

Microbiological Safety of Food, Part II. Report of the Committee on the Microbiological Safety of Food. Chairman: Sir Mark Richmond. HM Stationery Office, London, 1991.

Certain topics covered in specific chapters are dealt with in the following publications:

Chapters 2, 3 and 4

Doyle, M.P. (eds) (1989) *Foodborne Bacterial Pathogens*, Marcel Dekker, New York.

Finegold, S.M. and George, W.L. (eds) (1989) *Anaerobic Infections in Humans*, Academic Press, San Diego.

Willis, A.T. and Phillips, K.D. (1988) *Anaerobic Infections: Clinical and Laboratory Practice*, Public Health Laboratory Service, London.

Chapter 5

Betina, V. (1989) *Mycotoxins, Chemical, Biological and Environmental Aspects*, Elsevier, Amsterdam.

Krogh, P. (ed.) (1987) *Mycotoxins in Food*, Academic Press, London.

Kurata, H. and Ueno, Y. (eds) (1984) *Toxigenic Fungi – their Toxins and Health Hazard*, Elsevier, Amsterdam.

Marasas, W.F.O. and Nelson, P.E. (1987) *Mycotoxicology*. Pennsylvania State University Press, University Park.

Chapter 7

Tenover, F.C. (ed.) (1989) *DNA Probes for Infectious Diseases*, CRC Press, Boca Raton, Florida.

McPherson, M.J., Quirke, P. and Taylor, G.R. (eds) (1992) *PCR: A Practical Approach*, Oxford University Press.

Chapter 8

Farmer, R. and Miller, D. (1991) *Lecture Notes on Epidemiology and Public Health Medicine*, 3rd edn, Blackwell Scientific Publications, Oxford.

WHO Surveillance Programme for Control of Foodborne Infections and Intoxications in Europe. 5th Report published by FAO/WHO Collaborating Centre for Research and Training in Food Hygiene and Zoonoses, Robert von Ostertag Institute, Berlin.

Interim Report on Campylobacter. Advisory Committee on the Microbiological Safety of Food. HM Stationery Office, London, 1993.

Sibbald, C.J., Fitzsimmons, D. and Upton, P.A. (1994) *Communicable Diseases Scotland Weekly Report*, 28, no. 2.

Brown, P., Kidd, D., Riordan, T. (1988) An outbreak of food-borne *Campylobacter jejuni* infection and the possible role of cross-contamination. *Journal of Infection*, **17**, 171–176.

Bannister, B.A. (1987) *Listeria monocytogenes* meningitis associated with eating soft cheese. *Journal of Infection*, **15**, 165–168.

Davis, M., Osaki, C., Gordon, D. *et al.* (1993) Update: Multistate outbreak of *Escherichia coli* 0157:H7 infections from hamburgers – Western United States, 1992–1993. *Morbidity and Mortality Weekly Report*, **42**, (14), 258–263

Arnold, D., Litchfield, P. and Walker, D. (1989) An outbreak of food poisoning from frozen oysters. *Communicable Diseases Scotland Weekly Report*, **23**(9), 5–7.

Chapter 9

Modelling microbial growth and survival in relation to food safety and stability caught the attention of food microbiologists, resulting in a substantial scientific literature. Several useful reviews have been published (McMeekin *et al.*, 1992; Buchanan, 1993; Labuza and Fu, 1993; Foss and McMeekin, 1994; Skinner and Larkin, 1994; Whiting and Buchanan, 1994). The proceedings of the first international conference on predictive microbiology were published as a special number of the Journal of Industrial Microbiology (1993) (see Labuza and Fu (1993) as an example). A Special Issue of the International Journal of Food Microbiology (Farkas, 1994) contained five reviews and eleven research papers. The first book on predictive microbiology (McMeekin *et al.*, 1992) is essential reading for those wishing to address the mathematics. Due attention should also be given to the mathematical framework and the biological basis of models (Baranyi and Roberts, 1994, 1995).

Baranyi, J. and Roberts, T.A. (1994) A dynamic approach to predicting bacterial growth in food. *International Journal of Food Microbiology*, **23**, 277–294.

Baranyi, J. and Roberts, T.A. (1995) Mathematics of predictive microbiology. *International Journal of Food Microbiology*, **26**, 199–218.

Buchanan, R.L. (1993) Predictive food microbiology. *Trends in Food Science and Technology*, **4**, 6–11.

Farkas J. (ed) (1994) Special Issue: Predictive Modelling. *International Journal of Food Microbiology*, **23**, 241–477.

International Commission on Microbiological Specifications for Foods (1980a) *Micro-organisms in Foods. 3. Microbial Ecology of Foods. Vol. 1, Factors Affecting Life and Death of Micro-organisms*, Academic Press, New York.

International Commission on Microbiological Specifications for Foods (1980b) *Micro-organisms in Foods 3. Microbial Ecology of Foods, Vol. 2, Food Commodities*, Academic Press, New York.

International Commission on Microbiological Specifications for Foods (1986) *Micro-organisms in Foods. 2. Sampling for Microbiological Analysis: Principles and Specific Applications*, 2nd edn, University of Toronto Press, Toronto.

International Commission on Microbiological Specifications for Foods (1988) *Micro-organisms in Foods. 4. Applications of the Hazard Analysis Critical Control Point (HACCP) System to Ensure Microbiological Safety and Quality*, Blackwell Scientific Publications, Oxford.

International Commission on Microbiological Specifications for Foods (1994a) *Choice of sampling plan and criteria for Listeria monocytogenes. International Journal of Food Microbiology*, **22**, 89–96.

International Commission on Microbiological Specifications for Foods (1994b). *Micro-organisms in Foods 5. Characteristics of Microbial Pathogens.* Chapman & Hall, London.

Labuza, T.P. and Fu, B. (1993) Growth kinetics for shelf-life prediction: theory and practice. *Journal of Industrial Microbiology*, **12**, 309–323.

McMeekin, T.A., Ross, T. and Olley, J. (1992) Application of predictive microbiology to assure the quality and safety of fish and fish products (review). *International Journal of Food Microbiology*, **15**, 13–32.

McMeekin, T.A., Olley, J.N., Ross, T. and Ratkowsky, D.A. (1993) *Predictive Microbiology*, John Wiley & Sons Ltd, Chichester, UK.

Ross, T. and McMeekin, T.A. (1994) Predictive microbiology: *International Journal of Food Microbiology*, **23**, 241–264.
Skinner, G.E. and Larkin, J.W. (1994) Mathematical modelling of microbial growth: a review. *Journal of Food Safety*, **14**, 175–217.
van Impe, J.F., Nicolai, B.M., Martens, T. Baerdemaeker, J. and Vandewalle, J. (1992) Dynamic mathematical model to predict microbial growth and inactivation during food processing. *Appl. Environ. Microbiol.*, **58**, 2901–2909
Whiting, R.C. and Buchanan, R.L. (1994) Microbial modelling. *Food Technology*, **48**(6), 113–120.

Chapter 10

Jukes, D.J. (1993) *Food Legislation of the UK: A Concise Guide*, Butterworth-Heinemann, Oxford.

Chapter 11

Sprenger, R.A. (1993) *Hygiene for Management: A text for food hygiene courses*, Highfield Publications, Doncaster.

In addition to textbooks there are a number of good journals which often include relevant articles, including:

> *British Food Journal*
> *Epidemiology and Infection*
> *International Journal of Food Microbiology*
> *Journal of Applied Bacteriology*
> *Journal of Clinical Microbiology*
> *Journal of Food Protection*
> *Journal of Food Technology*
> *Journal of Infectious Diseases*
> *Journal of Medical Microbiology*
> *The Lancet*

National data and other relevant epidemiological information are published regularly in the PHLS *Communicable Disease Report* (CDR), the *Scottish Centre for Infection and Environmental Health (SCIEH) Weekly Report* and the *Northern Ireland Communicable Diseases Monthly Report*.

Index

Page number appearing in bold refer to figures and page number appearing in *italic* refer to tables.